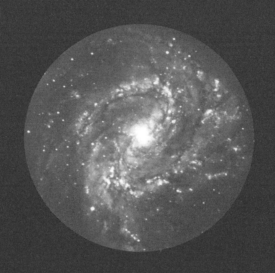

GOOD SEEING

GOOD SEEING

A Century of Science at the

Carnegie Institution of Washington

1902-2002

James Trefil

and

Margaret Hindle Hazen

With a Foreword by Timothy Ferris

JOSEPH HENRY PRESS
Washington, D.C.

Lower section of the 100-inch telescope at Mount Wilson placed in the fork, ca. 1916.

Frontispiece: Globular clusters, such as M80, which is 28,000 light-years from Earth, contain hundreds of thousands of ancient stars bound by mutual gravitation. This image is from the Hubble Space Telescope.

Front endleaf: Galaxy NGC 5236, photographed by Edwin Hubble on March 2–3, 1925.

Back endleaf: From Nettie Stevens's Studies in Spermatogenesis, *1905-1906.*

FOREWORD

*S*cience has revolutionized human life in ways as conspicuous as the rising sun. Scientific technology has rescued millions from hunger, want, and an early grave, producing advances in communications and transportation that have brought the world's peoples closer together. Scientific knowledge has transformed our universe, proffering radically improved conceptions of its evolution and of our own.

Less conspicuously, but in the long run even more importantly, the scientific method has provided humankind with a new and uniquely powerful method of inquiry. Its rise is routinely described as a triumph of reason over superstition, but this truism misses the point. For thousands of years prior to the advent of science, philosophers constructed eminently reasonable, logical systems that purported to explain just about everything. Compared to the grandeur of these philosophical constructions, science seemed limited and tentative.

Scientists, as they later came to be called, were more often mistaken than correct in their ideas about nature—and were obliged with dismaying regularity to admit it. They stood mute in the face of the big questions like "What is beauty?" and "What is the purpose of life?" They spent much of their time tinkering with obscure machinery and foul-smelling chemicals, exploring questions so obscure as to be unclear even to themselves. Yet when it came to understanding how nature works, predicting its behavior, and harnessing its vast power, this halting, relatively unassuming approach produced deeper, wider, and more consequential revelations than had been attained by all the previous endeavors combined. To paraphrase the epitaph of the architect Sir Christopher Wren, if you seek a monument to science, look around.

But the revolution occurred so recently—science has been a going concern for only about four hundred years—that it has not yet fully penetrated the consciousness of even those cultures most involved with it. As the astronomer Carl Sagan observed, "We live in a society exquisitely dependent on science and technology, in which hardly anyone knows anything about science and technology." The result is a dangerous dichotomy between an involuntary elite who are conversant with science and understand it to be a force for individual liberation and an undereducated majority, misled by pop-culture exploitations of the Faust myth, to whom science is—as the French say of the law—a machine that cannot move without crushing someone.

Alfred G. Mayer at the wheel of the Carnegie yacht Physalia, September 5, 1904.

Nowhere is the danger of popular estrangement more evident than in the management of scientific research. Those who fail to understand science—and their ranks include, unfortunately, most of the political leadership of the industrialized world—tend to think of it as an instrument of power, a kind of ray gun that can blast problems into smithereens. From this misconception arise blundering crash programs like the ill-conceived "war on cancer." When concerned mainly with practical applications, such efforts can bear fruit—the Manhattan project, the Apollo program, and the Human Genome Initiative come to mind—but cutting-edge scientific research seldom works that way. Scientific inquiry in action is not a lean, clean laser beam but a jumble of fumbling guesses and messy experiments, most of them destined to fail. Yet experience shows that it is in this seemingly underorganized, undirected fashion that we most often actually get somewhere. As Winston Churchill said of democracy, science is the worst possible system, except for all the others.

The reasons for the near-anarchy of the scientific enterprise are not difficult to discern. Any first-rate scientist is engaged in grappling with the unknown: As we read in the present volume, "Scientists build instruments and do experiments not because of what we know, but because of what we don't know." Any such situation demands creativity, which is inherently unpredictable. Creative scientists don't know where they're going any more than creative artists do, and anyone who tries to direct research without understanding that fact is destined to share the frustrations of moneyed patrons who shouted themselves red in the face trying to direct and constrain the labors of Michelangelo, Beethoven, or Frank Lloyd Wright.

It was central to the genius of Andrew Carnegie that he understood how science actually works and acted on his understanding. The sort of man who thrived on the scientific and technological revolution, Carnegie rose from humble origins as a bobbin boy and messenger to amass a fortune based in large measure on his grasp of the process of empirical creativity: It is often remarked that he was the first industrialist to routinely employ scientists, but it is less often appreciated that he let them work without jiggling their elbows. By the time he founded the Carnegie Institution he was one of the most powerful men in the world, who might forgivably have fashioned the Institution into a tool of personal will, to be directed at solving the problems that most concerned him.

He did next to none of that. Instead, he chartered it to foster the pursuit of scientific knowledge for its own sake, hoping that such research would help with "the improvement of mankind" but understanding full well that neither he nor anyone else could predict where it would lead. As John Shaw Billings, one of the Institution's most influential trustees, remarked, "Experimental investigation . . . [is] uncertain in its results [which] will probably not be what we now expect." Vannevar Bush, celebrating the Institution's 50th anniversary, declared "that the Institution's primary responsibility is to carry on fundamental research in science with no initial view to its industrial application," and that only secondarily need it "explore some of the broader possibilities of use for human betterment."

The popular media of Carnegie's day understood him little better than they did science. Journalists called him a "robber baron"—a term borrowed from the dying age of aristocracy, employed to describe landowners who stole from travelers trespassing on their estates. The myth arose that the public libraries Carnegie donated were required to hang his portrait on the wall, as one would expect of a royal patron. Actually, he stipulated only that his libraries bear an image of the sun, with the maxim, "Let there be light." He wanted people to look, not at his pointing finger, but at where it was pointing—toward a potential future in which scientific creativity had banished superstition and fear. More modern than his critics, Carnegie understood that science is fundamentally democratic and egalitarian in nature, in that its fortunes depend on the education of the many, not the guidance of the few. If he saw the future more clearly than most, it was because he better appreciated how poorly we can predict the future of a scientifically and technologically creative world.

In the subsequent century of its illustrious career, the Institution has embodied and exemplified the spirit of scientific research. Mistakes were made, projects launched with high hopes sometimes had later to be abandoned, and sponsored research often proceeded haltingly, with little of the aura of gleaming inevitability that had graced the priestly and philosophical pronouncements of old. Yet out of this jumble arose knowledge of lasting value, about subjects as diverse as the behavior of the earth's magnetic field, the dynamics of earthquakes, the origin of life, the growth of plants, the spread of disease, the development of the human embryo, the evolution of stars, and the march of far-flung galaxies. Lives were saved and minds opened.

Now, as the Carnegie Institution embarks on its second century, the challenge of navigating the future has become more acute than ever. As the present authors put it, the scientific endeavor calls for "good seeing." To astronomers, "good seeing" means steady air, through which they can see clearly and far. Astronomers peer into the past: Looking at the Andromeda galaxy, two million light years from Earth, they see that vast city of stars as it was two million years ago, when *Homo habilis*, the first true human, appeared on Earth. Peering into the farthest reaches of space, they see the universe in its infancy. The point of studying the past is, however, to understand the present and anticipate the future. So this volume is commended to its readers, in the hope that they may have clear skies, "good seeing," and the liberty to explore and improve the surprising world that Mr. Carnegie helped bring to light.

TIMOTHY FERRIS

ACKNOWLEDGMENTS

This book would not have been possible without the help of many people. We are extremely grateful to the five scientists who served on the Carnegie Institution's Centennial Steering Committee: Louis Brown, Joseph Gall, Robert Hazen, George Preston, and Shauna Somerville. They were unfailingly generous in sharing their time and expertise. Owing to his close ties to the authors, Robert Hazen made contributions that surpassed all reasonable expectations. This book could not have been written without him.

We are indebted to the directors of the Carnegie departments: Wesley T. Huntress, Jr. (Geophysical Laboratory); Augustus Oemler, Jr. (the Observatories); Sean C. Solomon (Department of Terrestrial Magnetism); Christopher Somerville (Department of Plant Biology); and Allan C. Spradling (Department of Embryology). All helped with this project at various points and in various ways. We are also indebted to Vera Rubin, who graciously agreed to be interviewed and who provided imagery for her chapter.

A network of archivists and researchers enhanced the content of the book. Special thanks go to Shaun Hardy and Merri Wolf (librarians, Broad Branch Road campus), John Strom (archivist, the Administration Building), and Linda Schweizer (Director of External Affairs, the Observatories) for their efforts on our behalf. Many other specialists were central to the writing process. These researchers include Clare Bunce (archivist, Cold Spring Harbor Laboratory), Sarah Demb and Nasrin Rohani (Peabody Museum of Archaeology & Ethnology, Harvard University), Judith Goodstein and Shelley Erwin (archivists, California Institute of Technology), Nancy Langford (archivist, Dudley Observatory), and Dan Lewis (Curator of American Historical Manuscripts, The Huntington Library). Paul Ruther's unflagging assistance with imagery, fact-checking, and general research throughout the production of the book is appreciated more than we can say.

Thanks to Randolph Widmer, who provided useful comments for the archeology chapter. Thoughtful reviews of sections of the book by the following people also improved the text: Philip Abelson, Winslow Briggs, Donald Brown, Patricia Craig, Elizabeth Hanson, Nina Fedoroff, W. Kent Ford, Wendy Freedman, Ho-Kwang Mao, Adrianne Noe, Selwyn Sacks, Hatten S. Yoder, Jr., and George Wise. Any errors that remain, of course, are our own.

Stephen Mautner of the Joseph Henry Press provided insightful editorial guidance throughout the production phase of the book. We are grateful to him and his associate, Francesca Moghari, for producing such a handsome volume.

Most of all, we would like to thank Maxine F. Singer, who contributed to almost every phase of the project. Her ongoing willingness to share her knowledge—both of science and this institution—have made this a better book. We are deeply grateful.

JAMES TREFIL
MARGARET HINDLE HAZEN

Carnegie Institution of Washington's administration building, Washington, D.C., 1998.

Following page:
The administration building nearing completion, ca. 1909.

PART ONE

BUILDING AN INSTITUTION

*I*n 1903, the Wright brothers flew a motorized airplane over the sands of Cape Hatteras, Jack London published *The Call of the Wild*, and Thomas Hunt Morgan championed a grant application to the Carnegie Institution by a newly-minted Ph.D. student named Nettie Maria Stevens.

The Carnegie Institution was just one year old at the time, and its funding procedures were still in the formative phase. Yet the very next year, the Institution awarded Stevens $1,000. The grant was renewed in 1905—the same year in which Stevens produced part one of a remarkable paper: "Studies in Spermatogenesis with Especial Reference to the Accessory Chromosome." This work, which was published by the Carnegie Institution, offered cytological evidence demonstrating that the X and Y chromosomes were associated with sex determination.

The report was revolutionary. Prior to Stevens's work, many scientists speculated that gender was determined by environmental factors such as food and temperature. Stevens's research helped change that view. Her work is also cited as the first demonstration of an inherited trait associated with a chromosome. Although the discovery was made simultaneously by Edmund Beecher Wilson at Columbia University, Stevens's contributions are indisputable. During her lifetime, she would publish 40 papers on such topics as chromosomes, regeneration, and taxonomy.

It is easy to imagine what Carnegie grant #177 meant to Stevens in 1904 and 1905. It has been said that "financial worries preoccupied her," and surely Carnegie money helped assuage some of those worries. Carnegie support must also have validated her work as a scientist. Nettie Stevens's career path was atypical. Born in Vermont in 1861, she attended Westfield Normal School in Massachusetts, after which she taught for a while. Then, at age 35, she traveled across the country and enrolled as an undergraduate at Stanford University. She went on to study biology at Bryn Mawr College, receiving her Ph.D. at age 42.

Nettie Stevens died of breast cancer in 1912, just nine years into a promising research career. But the Institution that funded her early work thrived. It would go on to support many other capable people, including her Bryn Mawr mentor, Thomas Hunt Morgan.

What sort of a place was this?

How is it possible that Andrew Carnegie, pragmatic maker-of-steel, managed to create such a dynamic foundry of original thought?

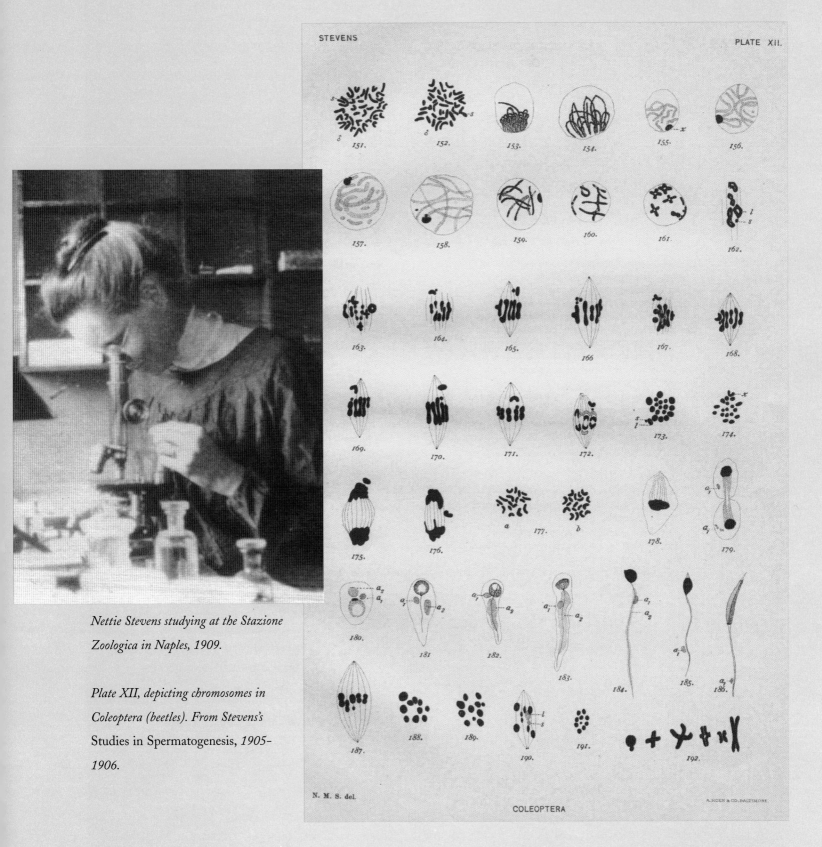

Nettie Stevens studying at the Stazione Zoologica in Naples, 1909.

Plate XII, depicting chromosomes in Coleoptera (beetles). From Stevens's Studies in Spermatogenesis, *1905–1906.*

Carnegie Libraries

During his lifetime, Carnegie built almost 2,000 public libraries in the United States and another 865 worldwide. A firm believer in education, Carnegie viewed libraries as ideal vehicles for the dissemination of knowledge. But Carnegie was also a firm believer in self-help, so he gave money for the buildings only. The books were the responsibility of the community.

Postcards depicting Carnegie-funded public libraries.
Clockwise, from top left, Hornell, New York; Shawnee, Oklahoma; Atlanta, Georgia; Madison, Minnesota; Washington, D.C.; Coffeyville, Kansas; and Flint, Michigan.

CHARTING THE COURSE

During the summer of 1901, Andrew Carnegie decided to establish an institution of higher learning in the nation's capital. As was his custom, he listened to the advice of experts—in particular, Daniel Coit Gilman, the recently retired president of the Johns Hopkins University, and Andrew D. White, president emeritus of Cornell University, both of whom favored funding a university in George Washington's honor. George Washington himself had suggested the creation of such a school when the federal city was on the drawing boards, and now, some 100 years after the first president's death in 1799, there was strong interest in reviving the memorial project.

Carnegie decided against it. He believed that establishing a top-ranked university would require much more money than he was willing to spend. Moreover, he didn't want to compete with existing universities. He preferred to explore what historian of science George Sarton would later call "untrodden paths." This approach led Carnegie to the creation of a riskier, and far more exciting, enterprise: an independent research organization that would support investigation and discovery simply for the sake of increasing knowledge.

The key meeting in the founding of the Carnegie Institution occurred on November 16, 1901, at Carnegie's recently completed mansion on East 91st Street in New York. Present at this meeting was Daniel Gilman, whom Carnegie immediately tapped for the presidency of the fledgling institution. Also present was John Shaw Billings, founder of the Library of the Surgeon-General's Office (now the National Library of Medicine), director of the New York Public Library, and one of the most important figures in the early years of the Carnegie Institution. Billings, a brilliant surgeon who had served at Gettysburg and Chancellorsville during the Civil War, had gone on to conduct scientific research on cattle disease and other issues related to public and military health. Billings lobbied forcefully for a research institute as opposed to a school. He also enthusiastically endorsed the idea that the institution should support "exceptional" people in their life's work.

These concepts weren't original with Billings. John D. Rockefeller had funded a medical research institute in New York a few months earlier, and Alfred Nobel's annual awards for individual excellence in a range of disciplines began during the same year. But both concepts resonated strongly with Andrew Carnegie. Carnegie was a man whose respect for "geniuses" bordered on reverence.

John Shaw Billings, 1913.

Moreover, he had received an intriguing letter from Arthur Balfour, prime minister of England, the previous summer. In this letter, Balfour strongly advocated the establishment of enterprises that increased the world's stock of knowledge in addition to those that merely disseminated such knowledge. Carnegie remembered Balfour's advice and took it seriously. In fact, by the end of that November meeting in New York, only the details of Carnegie's institution of discovery remained to be worked out.

Over the next six weeks, the foundation for the Carnegie Institution was set—if not in stone, at least on paper. On January 4, 1902, the articles of incorporation were discussed, approved, and filed with the District of Columbia recorder of deeds. (The official name on this seminal document—Carnegie Institution— gave way in 1904 to Carnegie Institution of Washington.) Also at this meeting, 27 distinguished trustees were selected by a protoexecutive committee that included Gilman, Billings, and several others central to the development of the Institution. Andrew Carnegie was especially pleased with the endorsement of President Theodore Roosevelt, who agreed to serve as an ex-officio trustee. Other eminent men from various walks of life agreed to serve more actively. These included Elihu Root (secretary of war); Seth Low (mayor of New York); William N. Frew (president of the Carnegie Institute of Pittsburgh); Charles D. Walcott (director of the U.S. Geological Survey); S. Weir Mitchell (physician and novelist); John Hay (secretary of state); and Carroll D. Wright (U.S. commissioner of labor). In a letter dated November 18, 1901, Billings championed the inclusion of Mrs. Henry Draper because he thought it "probable that it will be best to have one woman on the larger Board." Mary Anna Draper, the widow of a gifted amateur astronomer, didn't make the final cut (it wasn't until Margaret Carnegie Miller joined the board in 1955 that a woman was invited to serve as a trustee), but the nature and "extraordinary high quality" of the list pleased Andrew Carnegie enormously.

By mid-December, news of the project and its purpose had spread far and wide—though not always to rave reviews. "Some may oppose gift," read a headline on December 11, 1901. Part of the opposition came from the enormity of the donation. A $10 million endowment, which was what Carnegie proposed, was huge by the day's standards. It equaled Harvard's endowment and far exceeded the research monies in American universities as a whole. Some people worried that the concentration of such wealth in a single institution was unwise. Others viewed Carnegie's gift, which was in steel bonds, as tainted money.

But planning continued. The deed of trust was drafted, amended, and finalized. Then, toward the end of January, all was in place. At 2:30 pm on January 29, 1902, the Institution's first board of trustees gathered at the Department of State to receive their charge from the founder. Andrew Carnegie presented his deed of trust along with his good wishes. "Gentlemen," he said, "Your aims are high, you seek to expand known forces, to discover and utilize unknown forces for the benefit of man. Than this there can scarcely be a greater work . . . I wish you abundant success."

The gift—as well as the responsibilities that it entailed—was unanimously accepted by the board the next day. The work of the Institution was poised to begin.

First meeting of the board of trustees, January 29, 1902.

Left to right, standing: Carroll D. Wright, S. Weir Mitchell, S. P. Langley, D. O. Mills, John S. Billings, Abram S. Hewitt, William W. Morrow, Wayne MacVeagh, Charles L. Hutchinson, William N. Frew, Daniel C. Gilman, Henry L. Higginson, Henry Hitchcock, Charles D. Walcott, William Lindsay.

Left to right, seated: D. B. Henderson, William P. Frye, Lyman J. Gage, Elihu Root, John Hay, Andrew Carnegie.

"Gentlemen, Your Work Now Begins"

But how were they to begin in a practical sense? The deed of trust contained a simple, two-part statement of purpose: (1) to encourage investigation, research, and discovery "in the broadest and most liberal manner," and (2) to foster "the application of knowledge to the improvement of mankind." Beyond this commitment to what would now be termed "pure" research and the almost predictable nod to the importance of pursuing practical applications of discoveries, there were few specifics in the Institution's founding documents. Stated outright was Carnegie's fervent hope that the new organization would secure American leadership in the world of scientific research. Unstated were instructions on how, in

a workaday sense, the managers of the new institution might disburse funds in pursuit of these stated goals. Carnegie clearly hoped that the managers of the Institution would find a way to implement what the original deed of trust specified only vaguely, namely, "to discover the exceptional man in every department of study whenever and wherever found . . . and enable him to make the work for which he seems especially designed his life work." Support of such exceptional people could extend, if necessary, to the provision of "such buildings, laboratories, books, and apparatus, as may be needed." Beyond these provisos, however, there were few specifics to guide the Institution's founding officers.

"Sweeping . . . in scope yet obscure in . . . particulars." That is how historian of science Nathan Reingold characterizes the Institution's founding papers, and that is exactly how Andrew Carnegie planned it. On presenting the deed of trust, he posed a rhetorical question to his board: "Have you ever seen or heard of a body of men wise enough to legislate for the next generation?" Carnegie asked. "No," was the expected answer and that, Carnegie went on to explain, was precisely why he had inserted the all-important "last clause." This clause stated that with a two-thirds majority the trustees could change funding practices according to the "changed conditions of the time." Such freedom proved to be an advantage over time. It enabled the Institution to adapt its work to times of war, just as it enabled it to adjust its programs to the vagaries of the economics of the larger world. More important, the elasticity of the deed encouraged the pursuit of questions, not bureaucracy. It virtually demanded that research remain forward looking and vital.

In the short run, of course, such vagueness posed a challenge. The Institution's second president, Robert Woodward, compared the process of defining the Institution's mission to the struggle of an organism "trying at once to discover its proper functions and to adjust itself to the conditions of its environment." But difficult as it was, the process of discovery had to begin somewhere, and not surprisingly, it began in the institutional board room with a series of administrative decisions. On the day of the presentation of the deed, Daniel Gilman was elected president and Charles D. Walcott was elected secretary. The next day, an executive committee was formed to carry out the business of the Institution. In addition to Gilman and Walcott, this executive committee included John Shaw Billings, Elihu Root, Abram S. Hewitt, S. Weir Mitchell, and Carroll D. Wright. The executive committee was pivotal to the shaping of the Institution in its earliest days. Indeed, an elite subset of the committee, Billings and Walcott, virtually ran the Institution in its early years.

One of the first acts of the executive committee was to create 18 advisory committees to solicit guidance from experts about the important questions of the day. The range of fields was wide: botany, economics, physics, geology, geophysics, geography, meteorology, chemistry, astronomy, paleontology, zoology, physiology, anthropology, bibliography, engineering, psychology, history, and mathematics. The reports of these committees were published in the Institution's first *Year Book*, and they had great impact on subsequent funding decisions. In fact, eight advisers saw their own large-scale dream projects transformed into reality with Carnegie money. Many more received minor grants in their particular fields of interest.

Daniel C. Gilman, Carnegie Institution president, 1902–1904.

SCIENCE

NEW SERIES.
VOL. XVI. No. 416.

FRIDAY, DECEMBER 19, 1902.

SINGLE COPIES, 15 CTS.
ANNUAL SUBSCRIPTION, $5.0

Dear Mr. Carnegie . . .

Scientists from across the country responded to James McKeen Cattell's call for advice on how to spend Mr. Carnegie's money. Some scientists expressed fear that the enormous endowment might affect scientific research adversely, but most saw a golden opportunity for funding their own pet projects.

"I trust that the trustees of the Carnegie Institution will make it a point to cooperate with the heads of our great museums in preserving for the students of American science the types of all American species."

W. J. Holland, Carnegie Museum, Pittsburgh, September 23, 1902

"I should like to see at Washington a Carnegie Institution somewhat on the plan of the Royal Institution of London, which, as we all know, was founded by an American . . . . It should contain comparatively small but admirably equipped laboratories for the three fundamental sciences—physics, chemistry, and psychology."

James McKeen Cattell, editor of *Science*, September 19, 1902

"It is not so much 'what should be done' as 'what is not doing.' Lines of research which are not now followed should receive first attention. In every branch of science is found a great unknown, an unexplored desert. Into these regions scientific research should penetrate."

H. W. Wiley, U.S. Department of Agriculture, September 26, 1902

"The establishment of laboratories at Washington for special investigations not yet well provided for seems to me most legitimate. One example of such an institution would be a breeding house or vivarium for the study of heredity."

D. S. Jordan, October 3, 1902

"All workers in science need skilful and energetic help in the thankless drudgery of reference hunting. To give them the necessary aid the Institution should at once subsidize the Concilium Bibliographicum."

Bashford Dean, Columbia University, October 24, 1902

"I would respectfully submit that a portion of the income might well be made available to the smaller, but poorly endowed, colleges to enjoy the advantages of a sabbatical year."

C. W. Stiles, U.S. Public Health Service, October 31, 1902

Advice came from another source: the readers of *Science* magazine. Carnegie's grant was so large and had such potential for influence on American research that James McKeen Cattell, editor of *Science,* invited readers to submit their opinions about fruitful avenues of study. Between September 19 and December 26, 1902, Cattell published nearly 50 letters in what was, essentially, a public debate on the mission of the Carnegie Institution.

Long before the *Science* magazine opinion poll was launched, the first Carnegie grants had been made. In April 1902, the Woods Hole Marine Biological Laboratory received $4,000 for general support. (There was talk of the Carnegie Institution taking control of the entire laboratory, though this idea proved so unpopular with Woods Hole scientists and others that it was eventually dropped.) The first individual grants were also made in April 1902. James McKeen Cattell of Columbia University received $1,000 to prepare a list of American scientists. The same amount was awarded to Dr. Hideyo Noguchi and Professor Simon Flexner to support their studies of snake venom.

In retrospect, it is clear that these three grants set the pattern of Carnegie funding for many years. That is, the Institution would support both individual researchers and "larger projects." The only question, as historian Robert Kohler has pointed out, was in what proportion and to whom.

Individual Grants

Initially, almost inevitably, the trustees favored grants to individuals because such grants were easier and safer to fund than the larger ones. In addition, many on the board favored individual grants because they seemed to jibe most closely with the founder's vision.

That didn't mean that determining the recipients of such grants was easy. Requests for money from self-proclaimed "exceptional" men had been pouring in from the moment the Institution was launched. By December 1903, the Institution had received 1,042 requests for money in 33 fields. Decisions had to be made, and quickly. Daniel Gilman, who preferred to "make haste slowly," began to be eclipsed by Billings and Walcott, both of whom had strong views on the running of an institution. In 1904, Gilman resigned, defeated by his colleagues' relentless zeal. By that time, however, many individual grants had been made. In 1903 and 1904, for instance, a total of $24,500 was granted to geologist–archeologist Raphael Pumpelly to study migration patterns of early populations in Eurasia, particularly the ancient site of Anau (in modern Turkmenistan). Most other grants were smaller: Albert Michelson at the University of Chicago received $1,500 for aid in making ruling diffraction gratings (which were used in astronomy); Oliver Hay of the American Museum of Natural History received $3,000 to study fossil turtles; Janet Perkins, working at the Royal Botanical Gardens in Berlin, received $1,900 for preliminary studies on Philippine flora. In these early years, the humanities were still a part (albeit a small one) of the Institution's mission. Thus, Ewald Flügel, a professor at Stanford University, received $7,500 to prepare a lexicon for the works of Chaucer.

In the years that followed, the Institution awarded scores of other individual grants, many with outstanding results. For instance, geneticist–embryologist Thomas Hunt Morgan and his "fly group" received support for their research on *Drosophila* genetics at Columbia University and at the California Institute of Technology from 1914 through 1945, three years after Morgan himself retired. Similarly, geologist T.C. Chamberlin received support for his studies of fundamental problems of geology at the University of Chicago from 1903 until 1928. Historian of science George Sarton was recruited by George Ellery Hale in 1918. He received Carnegie support for his writing and lecturing—most of which was done in Cambridge, Massachusetts—for the next 30 years.

The funding of such "lone geniuses" had drawbacks, however. The Institution had little control over the work, and there was occasionally a tendency toward isolationism that was counterproductive to good work. Nowhere were these difficulties more clearly exemplified than in the Institution's experience with Luther Burbank.

In 1902, no one doubted that Luther Burbank was an exceptional man, including Luther Burbank himself. Though he lacked an advanced education, Burbank had developed hundreds of new varieties of plants in his California nurseries. It appeared that he had the power to increase the world's food supply. He seemed the ideal recipient of a Carnegie research grant, and at first the Carnegie trustees agreed. In 1905, they awarded Burbank $10,000, thereby creating what amounted to a Department of Horticulture.

The hope was to discover the scientific underpinnings of Burbank's methods of hybridization and in the process to shed light on questions of heredity. To this end, the Carnegie Institution sent George Shull, a biologist at the Institution's recently formed Department of Experimental Evolution at Cold Spring Harbor

Luther Burbank, 1909.

George Shull.

on Long Island, to study the master's ways. Between 1906 and 1911, Shull made several trips to Santa Rosa, where he worked side by side with Burbank in his experimental gardens. Ultimately, however, Shull was baffled. The "genius" was reticent and his procedures were unscientific. Furthermore, as Shull dryly remarked, Burbank's success seemed to rest most heavily on "the correctness of his economic ideals."

The Carnegie board of trustees discussed the "Burbank problem" many times. Andrew Carnegie strongly endorsed Burbank's work and urged the Institution's continued support. "I think a man of science should go cap in hand to a genius," Carnegie said. The Burbank episode is, in fact, a rare instance of Andrew Carnegie's gentle intervention in the administration of the Institution.

(The only other issue Carnegie cared strongly about was simplified spelling, which he hoped would be adopted in all official documents.) Robert Woodward, who was president of the Institution at the time, held out hope that a scientific evaluation by Shull might legitimize the money spent on Burbank. But it wasn't meant to be. On December 14, 1909, the board voted to terminate involvement with Burbank.

The decision was an important one. It demonstrated that the Institution could stop unsuccessful ventures when necessary. More significantly, it under-scored the wisdom of Woodward's organizational vision, which he had been

Andrew Carnegie filling the coffers of the Carnegie Institution, by cartoonist Clifford Berryman, published in the Washington Evening Star, *January 1911.*

promoting since he assumed the Institution's presidency in 1904. The Institution should support professional people of proven abilities, Woodward insisted. It shouldn't support lone geniuses tinkering on the fringe.

Woodward has been called the first modern manager of science. He rejected the romantic nineteenth-century approach to research, and instead championed the collectivist–specialist approach that we recognize today. But this doesn't mean that Woodward didn't appreciate an "exceptional" scientist when he saw one. George Shull's research at the Department of Experimental Evolution established the principles for the creation of hybrid corn, thereby laying the groundwork for a worldwide revolution in agriculture. He was exactly the sort of scientist Woodward was looking for: an exceptional person working in a "managed" setting.

By the same token, Woodward was exactly the sort of person Andrew Carnegie was looking for to run his Institution. Unlike Daniel Coit Gilman, the gentlemanly and cautious first president, Woodward displayed a remarkable ability to chart a course for the fledgling Institution that made fiscal and intellectual sense. Carnegie and Woodward maintained a cordial working relationship until Carnegie's death in 1919 (Woodward retired the next year). Carnegie was so satisfied with the work of his institution of discovery under Woodward's leadership that he increased his original endowment of $10 million with gifts of $2 million in 1907 and $10 million more in 1911. These additional funds were paid largely because of Carnegie's satisfaction with the Institution's pursuit of "larger" projects, such as the building of telescopes at Mount Wilson and the launching of a sea observatory that would measure Earth's magnetic field.

Large Grants and Departments

In the early days, Andrew Carnegie had strong opinions about the establishment of institutionally sponsored laboratories and no reservations about expressing them: "You know my own opinion is that no big institution should be erected anywhere," Carnegie wrote to Walcott in 1905, "[E]xceptional men should be encouraged to do their exceptional work in their own environment." Then, to nail the point, "There is nothing so deadening as gathering together a staff in an institution. Dry rot begins and routine kills original work." Such was "the opinion of Yours very truly, Andrew Carnegie."

But Carnegie's views had already been overridden. As early as 1903, the board of trustees had authorized the establishment of three institutional research facilities: the Desert Botanical Laboratory in Tucson, the Station for Experimental Evolution on Long Island, and the Marine Biological Laboratory in the Florida Keys. It had also authorized the Department of Historical Research, which, though lacking a formal headquarters, received a mandate to promote historical inquiry. The following year, three more departments were authorized by the Institution: Terrestrial Magnetism, Economics and Sociology, and the Solar Observatory. By 1907, the Institution was funding 10 departments devoted to the study of the earth sciences, astronomy, biology, and history. It also funded an active publications program. And yet, the "dry rot" envisioned by Carnegie was rarely a problem. Why?

One answer is simply that "managed science" proved to be an excellent way to conduct science in the twentieth century. But that is only half an answer, of course. "Managed science" could be managed or mismanaged in any number of ways. To be effective, an institution requires oversight that is based on intellectual principles that work, principles that the Carnegie Institution espoused from the start—specifically, in the form of the rigorous criteria it imposed on its grantees. These criteria, in turn, owed a lot to John Shaw Billings.

Billings was often praised for his enthusiasm, for his courage in undertaking new enterprises: "Why, of course, we can do it," he is reported to have said. "Why else are we here?" But Billings was also a stern taskmaster. His passion for excellence, his preference for plain speech, and his insistence on fiscal responsibility led to a single conclusion: While the Carnegie Institution could develop in any number of ways, it couldn't develop in all ways simultaneously. Choices had to be made. But how?

On June 11, 1904, Billings attended the opening of the Carnegie Station for Experimental Evolution (later renamed the Department of Experimental Evolution) at Cold Spring Harbor on Long Island. Representing the Institution, Billings underscored the importance of the occasion by describing the Institution's requirements for support—individual and otherwise. "We have been in the habit of asking several questions," Billings said, regarding the never-ending stream of applicants for Carnegie money:

Is the proposed research likely to give good results?
Is this research being carried on elsewhere?
What are the qualifications of the investigator?
Does the project involve cooperation?

The only acceptable answers to these questions for Billings and the Institution as a whole were: yes; no; exceptional; and, most emphatically, yes.

In general, this has been the Carnegie philosophy all along, except for one additional point—time. Billings had something to say about time, too, on that warm June day: "We know that experimental investigation, especially in this field, is a slow process, and uncertain in its results, and that we must be patient. This is a seed that we are planting; for the buds and blossoms and fruits we must wait, believing that they will come in due season, although they will probably not be what we now expect."

Years later, Freeman Dyson characterized effective scientific research as "organized unpredictability." Billings's performance as a Carnegie trustee suggests that he would have heartily agreed.

CARNEGIE INSTITUTION OF WASHINGTON

WASHINGTON, D. C.

STATEMENT OF FUNDS REQUIRED

AND FUNDS ON HAND

Funds Required for Disbursement during ...Dec 24 — Jany 24. 1910

MONTHLY PAYMENTS:

Large Grants	$ 16636.76
Minor Grants	1,550.31
Research Associates and Assistants	1,258.48
Publications	543.26
Administration	7038.80
	27,027.61

ESTIMATED CALLS:

Dept. of Botanical Research	$ 5,000	
" Economics and Sociology	1,500	
" Experimental Evolution	12,000	
" Historical Research	2000	
" Marine Biology	1000	
" Meridian Astrometry	1000	
" Terrestrial Magnetism	6000	
Astrophysical Observatory	5000	
Geophysical Laboratory	2000	
Nutrition Laboratory	2000	
Publications	5000	
Administration building & equipment	20,000	67,500
		89,527.61

FUNDS ON HAND:

Checks ready for use	$35,000	
Balance in National City Bank	3,146.18	
Balance in American Security and		
Trust Company	4,377.68	42523.86
		47,003.75

Checks # 441 - 2 - 3 - 4 - 5 @ 10,000   50,000
440 - 6   "   5,000   10,000

THE DEPARTMENTS

From the outset, the Carnegie Institution has supported gifted researchers with inquiring minds. Not surprisingly, the questions asked by some of these investigators have been "big" questions: Why does Earth's magnetic field vary? How does life evolve from a single cell? What are the positions and motions of the stars?

But big questions require big resources—specifically, sophisticated equipment, cross-disciplinary collaborations, and the all-important safety net of an elastic time frame in which to get the work done. It was probably inevitable, therefore, that institutional departments would be formed to meet these big-ticket needs.

The Carnegie Institution has supported 11 departments at various times over the years. It has also maintained what amounts to an administrative "department" to oversee the Institution's work as well as the Publications Office—sometimes referred to as a department—to disseminate information to a wider audience.

It is immediately obvious from the following overview that there have been many "starts" and "stops" during the last century. The Department of Economics and Sociology, for instance, survived for just 12 years. The Department of Marine Biology lasted somewhat longer—from 1903 until 1939—before being closed down. Less obvious on the surface but equally important to the Institution's work are the starts and stops that occurred within individual departments. At midcentury, for instance, the Department of Embryology redefined its mission to take advantage of developments in molecular biology. Likewise, the Geophysical Laboratory welcomed biologists and chemists to its staff during the 1950s, thereby inaugurating a biogeochemical group that augmented the classic petrologic studies of the laboratory. The Department of Terrestrial Magnetism has reinvented itself so many times that renaming it now would insult the many scientists whose evolving visions shaped the department's research program over the years.

Some historians of science claim that in the world of scientific research "the work drives the work." To an extent, this is true. All professions are self-promoting. But in the world of Carnegie science, it is also true that questions drive the scientists, and they in turn drive the work.

Questions change over time. So does an institution.

The Marine Biological Laboratory at Tortugas, Florida, July 28, 1904. This image is the first photograph to be published in a Carnegie Year Book. The buildings were portable so they could be moved to a different site.

ADMINISTRATION

For several years, the Institution conducted business in rented quarters. However, the idea of erecting an administration building was discussed as early as January 30, 1902. Although Andrew Carnegie maintained that the Institution should be known for the grandeur of its work—not its architecture—the trustees authorized the construction of an administration building in 1905. They chose a prominent downtown site at the corner of 16th and P Streets and hired the distinguished firm of Carrère and Hastings, creators of the New York Public Library and two congressional office buildings, to provide an imposing design. The building was dedicated in 1909. Andrew Carnegie attended the ceremony and by all accounts was very pleased. An adjacent wing containing an auditorium and office space was added in 1938 to accommodate the Institution's public lectures and exhibitions.

As Andrew Carnegie knew, a building wasn't worth anything unless the activities that went on inside had merit. In this office building many decisions were made about the operation of the Institution, and as a result, about the operations of American science as a whole.

Prompt publication and distribution of research results were embraced as goals from the outset. The trustees allocated $40,000 for this purpose in 1902 alone. A year later, the Institution's first *Year Book* and first monograph, a survey of the Desert Laboratory, were published. Soon thereafter, publishing at the Institution was flourishing. By 1909, the Division of Publications had been authorized to handle the work. *Index Medicus* was just one aspect of this publishing initiative. The Institution published nearly 700 monographs on a variety of topics during its first 100 years. For years, these works were distributed free of charge to libraries across the country. In addition, until the last quarter of the twentieth century, the Carnegie *Year Books* functioned as professional journals. Related publications were printed in the Carnegie administration building's print shop.

In this day of digitized and instant information, it is easy to lose sight of the fact that getting and preserving information wasn't always easy. "Be sure to read promiscuously," Andrew Carnegie once said. He thought it good to learn.

The Carnegie Institution Presidents

DANIEL C. GILMAN
1902–1904

ROBERT S. WOODWARD
1904–1920

JOHN C. MERRIAM
1921–1938

VANNEVAR BUSH
1939–1955

CARYL P. HASKINS
1956–1971

PHILIP H. ABELSON
1971–1978

JAMES D. EBERT
1978–1987

EDWARD E. DAVID, JR.
acting president
1987–1988

MAXINE F. SINGER
1988–

*Detail of mural by J. Monroe Hewlett, commissioned
for the Institution's Elihu Root Auditorium, 1938.*

DEPARTMENT OF PLANT BIOLOGY
(1903–   )

Directors

Daniel T. MacDougal, Director of Desert Laboratory (1906–1927)

Herman A. Spoehr, Chairman, Division of Plant Biology (1928–1929; 1931–1947)

H. M. Hall, Acting Chairman (1929–1930)

C. Stacy French (1947–1973)

Winslow Briggs (1973–1993)

Joseph A. Berry, Acting Director (1993)

Christopher Somerville (1994–   )

The Carnegie Institution's Desert Botanical Laboratory, the first of its kind in the world, was established in 1903 outside Tucson, Arizona. Devoted to the study of desert plants (how they tolerated, adapted to, and interacted with their environment), the laboratory attracted a host of important American scientists, including Daniel T. MacDougal ("the battling botanist," a leader in the movement to put the study of evolution in a laboratory setting), Forrest Shreve (a gentle field ecologist from Maryland), and Frederic Clements (the controversial "father of ecology," whose ideas of belligerent succession of species were highly revered in his time). In 1910, Herman Spoehr joined the staff and inaugurated changes that were to have lasting repercussions. At the time of his arrival in Tucson, Spoehr was virtually the only student of photosynthesis in America. A charismatic leader, Spoehr would fully realize MacDougal's goal to set the department on a sound physiological course. Under Spoehr's direction and with the hearty approval of the Institution's president, John C. Merriam, the department reorganized and moved to Stanford, California, in 1929. There, on a sliver of Stanford University soil, the Carnegie Institution's plant biologists, now members of the Carnegie's "Division of Plant Biology," created what would soon become a world-class research organization.

Not surprisingly, the study of photosynthesis was to continue on an ever more sophisticated level. C. Stacy French (inventor of the "French pressure cell," used to study cellular structure and processes in plants), pioneered a biophysical approach to the work. His innovative experiments on pigments and pigment

Forrest Shreve using a dendrograph, which records fluctuations in plant circumference due to water content. Shown at the Carnegie Institution's Desert Laboratory, Tucson, Arizona, 1933.

complexes were extended by succeeding Carnegie researchers, including Harold Strain, David Fork, Arthur Grossman, and Winslow Briggs.

Other important research programs were also pursued at the department. In an extraordinary 20-year effort, the research team of Jens Clausen, William Hiesey, and David Keck provided powerful evidence about evolution in plants—specifically that, although changes in environment do not cause hereditary changes in plants, they do provide new habitats for species undergoing "natural" mutations. Other plant biologists, including Joe Berry and Chris Field, have built upon the earlier departmental efforts in ecology. Their studies of the long-term effects of carbon dioxide enrichment have provided useful information about greenhouse effects on a global scale. A direct application of their work is the formulation of government policy on reduction of greenhouse gas emissions.

Under the direction of Christopher Somerville since 1994, the department is now committed to using the tools of molecular biology to solve problems related to overpopulation, world hunger, and environmental degradation. In 2000, an international team of scientists, including Somerville, completed the sequencing of the *Arabidopsis* genome, the first for any plant. The information is invaluable for deducing the chemical mechanisms of plant development and the adaptations plants make to hostile and changing environments.

DEPARTMENT OF HISTORICAL RESEARCH
(1903–1958)

Directors

ANDREW C. MCLAUGHLIN (1903–1905)

J. FRANKLIN JAMESON (1905–1928)

W. G. LELAND, Acting Director (1929)

ALFRED V. KIDDER, Chairman (1929–1950)

H. E. D. POLLOCK, Director, Department of Archaeology (1951–1958)

Carnegie archeologists at Chichén Itzá, 1925 (seated, left to right): Oliver G. Ricketson, Edith Bayles Ricketson, Karl Ruppert, Earl H. Morris, Sylvanus G. Morley, Ann A. Morris, John C. Merriam (president of the Carnegie Institution).

It is a little-known fact that the original articles of incorporation of the Carnegie Institution called for the establishment of an organization that would promote "original research in science, literature, and art." Given the prominence of Gilman and other "humanitarians" on the early board, it is hardly surprising that "historical science" received funding from the start. But the ride would prove a bumpy one. Indeed, the history department fought for its life for over 50 years at the Carnegie Institution. That it failed to stay afloat says less about the Institution than the subject matter, which in the end was impossible to quantify in the ways that processes of the natural world could be quantified. One could pursue history with rigor, of course, but it remained essentially an art.

Not that the point mattered to some people. In 1909, the Institution received a petition from 10 national societies demanding increased funding for humanities across the board. Humanists, it seemed, were in a tizzy about being eclipsed by the hard sciences. They wanted a piece of the Carnegie pie.

In the early years, historians managed quite well. In December 1903, the trustees authorized $8,500 for the work of the Carnegie Bureau of Historical Research, and support continued well into the 1930s. Under the leadership first of Andrew McLaughlin and later of J. Franklin Jameson, this department embraced a simply stated mission: to help individual scholars engaged in historical research. In Jameson's view, "it must be the proper function of an organized and permanent institution, disposing of ampler resources than most individual historians can command, to carry on the primary, fundamental, and costly tasks of finding the materials or guiding men to them, and of printing such of them as are unprinted and most deserve print, selecting those which are likely to give the greatest possible aid and incitement to the production of good monographs in important fields." In other words, the department was not in the business of

writing histories but of helping others do so. To that end, many useful historical guides and documentary compilations were produced by Carnegie staff. These include *Letters of Members of the Continental Congress* (1933) and *Correspondence of Andrew Jackson* (1926–1929). Leo Stock's multivolume work, *Proceedings and Debates of the British Parliaments Respecting North America* (1924–1941), is an outstanding example of the sort of publication Carnegie scholars produced for the use of other scholars.

"Bonampak Mural, room 1, structure 1," from a Mayan site in Mexico. Painted by Antonio Tejada and published by the Carnegie Institution, 1955.

The department also served as a national clearinghouse for historians. Some of the ancillary services Jameson and his staff provided included advising graduate students on thesis topics, helping make arrangements for scholars to get access to foreign archives, and locating key documents for researchers. Throughout his tenure as director, Jameson also served as editor of the American Historical Association's quarterly journal. This was yet another way that the department—and the Institution as a whole—aided the cause of history during the first third of the twentieth century.

Jameson retired in 1928 to accept the chair of American history at the Library of Congress. Though his work at the Carnegie Institution would be carried on by others, the Department of History was thoroughly reorganized. Two other branches of historic inquiry, both of which had been supported by the Institution earlier, now came to prominence in this department.

Section of Aboriginal American History

The Carnegie Institution supported archeology from the beginning. In addition to several large grants made to Raphael Pumpelly to study civilizations in the Near East, the Institution made smaller grants to other investigators—$3,500 in 1904 alone. But when, in 1914, Sylvanus G. Morley persuaded the Institution to fund Mayan research in Central America, a large-scale and long-lasting archeological venture was launched. Over the years, the Institution would fund excavations at sites throughout Central America and the American Southwest. Carnegie scientists pursued a range of work. Some, like Morley, focused on Mayan hieroglyphics. Others, like Anna Shepard, focused on ceramic technologies. Still others, like Earl Morris and Karl Ruppert, excavated buildings and in many cases helped reconstruct them. Chichén Itzá is the most famous example of such work.

Alfred Kidder, who joined the Carnegie Institution in 1926, saw value in these and other projects. Yet he saw something more. It was Kidder's dream to draw on multiple disciplines in order to achieve a more complete understanding of ancient cultures. Under Kidder's leadership, biology, climatology, zoology, linguistics, physical anthropology, agriculture, and tropical medicine figured into the Institution's archeological research. Many useful publications emerged from these studies. In the end, however, the data overwhelmed the department. At Vannevar Bush's insistence, the department began to scale back, to cease collecting data and start synthesizing what had been learned thus far. Thorough synthesis never took place and the department was closed in 1958. But current archeological investigations continue to draw on the wealth of information produced by Carnegie archeologists.

Section of the History of Science

The History of Science division of the Institution owes its existence to one man, George Sarton. Institutional funding for Sarton's historical studies began in 1918 as an individual grant. By the time, the Department of History was reorganized in 1929, Sarton had acquired a following. Thus, in 1929, the Institution sponsored the work of Dr. E. A. Loew, a student of paleography, and W. A. Heidel, a student of philosophy.

But Sarton was the keystone of the section. His three-volume treatise on the history of science, the final volume of which was published by the Carnegie Institution in 1948, remains a classic in the field. The publication of *Isis,* the journal of the history of science, was subsidized by the Carnegie Institution for many years.

DEPARTMENT OF GENETICS
(1904–1971)

Directors

CHARLES B. DAVENPORT (1904–1934)

ALBERT F. BLAKESLEE (1935–1941)

MILISLAV DEMEREC (1941–1960)

BERWIND P. KAUFMANN, Acting Director (1960–1962)

ALFRED HERSHEY, Genetics Research Unit (1962–1971)

Charles Davenport was one of the earliest recipients of Carnegie Institution support. In 1904, he persuaded the trustees to establish the Station for Experimental Evolution at Cold Spring Harbor on Long Island, adjacent to the Brooklyn Institute's biological sciences laboratory. There, Davenport and others studied one of the most important questions of the day: How is genetic information passed from parent to child? The Cold Spring Harbor scientists studied many organisms,

Barbara McClintock examining corn samples, ca. 1963.

including corn, mice, chickens, and jimsonweed to answer this question. They also studied human traits, and that is where Davenport ran into trouble. He became a leader among eugenicists, who sought to use scientific principles of heredity to solve social problems. It was worse than bad science. It was bigotry shrouded in the cloak of scientific terminology.

This shameful initiative was ended by Vannevar Bush almost as soon as he became president of the Institution. Thereafter, the Department of Genetics entered a golden age of productivity. With the entrance of Milislav Demerec, who became director in 1941, research at the Department of Genetics moved to the forefront of a rapidly burgeoning field. With others, Demerec produced landmark work on the genetics of fruit flies. He also encouraged the formation of the "phage group," a coterie of gifted scientists who studied small viruses called bacteriophage. Out of this initiative came Alfred Hershey's Nobel Prize–winning work that established, once and for all, that DNA was the genetic material of life.

Barbara McClintock arrived at the Carnegie laboratory in 1942 and remained there for the rest of her life. Her research focused on maize, the same organism that George Shull had studied during the first decade of the century. McClintock's work took her deep inside kernels of corn, where she saw things that other people had never seen. Her discovery of transposons—genes that move—won her a Nobel Prize in 1983.

Owing to budgetary concerns, the activities of the department were scaled back during the mid-1960s. The department was closed in 1971. McClintock and Hershey continued to receive support until they retired.

DEPARTMENT OF MARINE BIOLOGY
(1904–1939)

Directors

ALFRED G. MAYER (1904–1922)

WILLIAM H. LONGLEY, Executive Director (1923–1937)

D. H. TENNENT, Executive Director (1938–1939)

At the turn of the twentieth century, the mysteries of ocean life captivated scientists across the globe. When the Carnegie Institution was founded, laboratories at Woods Hole, Massachusetts, and Naples, Italy, were actively engaged in research in this sphere. Some Carnegie advisers suggested that the Institution lend its support to these existing projects. Some experts went so far as to suggest that the Carnegie Institution take over the Woods Hole operation entirely. The advisory committee on zoology took exception to this idea. "While . . . cooperation should be the main feature of the Institution," they wrote in 1902, "the committee strongly indorse [*sic*] the establishment of a permanent [Carnegie-sponsored] biological laboratory as a central station for marine biology."

Alfred Mayer, a Harvard-trained zoologist who had worked with Alexander Agassiz, stood above other applicants for aid in this field. His proposal for a marine laboratory at the Dry Tortugas islands off the Florida Keys was approved by the board of trustees in December 1903. The U.S. Department of Commerce, which maintained a lighthouse at the site, issued a revocable license for these scientific operations. Thereafter, plans proceeded quickly.

Two portable laboratories and two small outbuildings were built in New York, then shipped to the Tortugas, where they were erected 1,000 feet north of the lighthouse in July 1904. (Portable buildings were used so that the Institution could move them to a different site, if desirable.) In addition, the facility boasted a windmill for pumping salt water and air, a dock, a cistern for rainwater, and 50 newly planted palm trees to provide shade and beautify the site. A 60-foot ketch-rigged yacht with a 20-horsepower auxiliary engine was also commissioned for research purposes. The yacht, *Physalia*, was constructed at East Boothbay, Maine, during the summer of 1904 at a cost of $6,037.60. Later the laboratory would acquire several more vessels: a gasoline-powered yacht, *Anton Dohrn*, and three launches: the *Velella*, the *Darwin*, and the *Bull Pup*.

Diver with 85-pound copper helmet at Dry Tortugas, Florida, site of Carnegie's Department of Marine Biology, ca. 1928.

The site—chosen for the purity of the ocean water, the proximity of the Gulf Stream, and the richness of life in nearby coral reefs—quickly became the best-equipped marine biological station in the tropical world. Mayer, who was based at Princeton, traveled to the field station regularly, arriving in the spring and remaining until the hurricane season drove him away in the fall. Over the years, some 150 investigators from the United States and abroad followed a similar pattern. Their studies ranged widely. Experimental work in ecology, heredity, regeneration, and growth as well as in-depth investigations of the geology, botany, zoology, and physiography of coral reefs are just some of the laboratory's projects. Between 1917 and 1920, four expeditions were undertaken by the department to study reefs in American Samoa. This work, which encompassed studies of geology, ecology, meteorology, and marine life, represents the first major study of a Pacific island with respect to its coral reefs.

The remoteness of this laboratory was its primary attraction. But remoteness was also a factor in the laboratory's demise. Hurricanes were a constant worry and, in 1910, a hurricane destroyed sizable portions of the facility. By 1939, the Institution concluded that, while the installation was well suited to summer research, its short season of operation and its distance from the mainland rendered its maintenance uneconomical. The facility was closed in 1939.

DEPARTMENT OF ECONOMICS AND SOCIOLOGY
(1904–1916)

Directors

CARROLL D. WRIGHT, Chairman (1904–1909)

HENRY W. FARNUM, Chairman (1909–1916)

The Department of Economics and Sociology was established in 1904 under the direction of Carroll D. Wright, secretary of labor and a Carnegie trustee. Wright's goal was to discover the laws that contributed to the success of the American system of production and government. To this end, he established 11 subdivisions of investigation, each to be overseen by an expert and staff based in various academic institutions and federal agencies around the country. This was, in effect, the Institution's first "virtual" department.

Strikers stop street car, New York, 1916.

The subdivisions were charged with an enormous task: to accumulate information on all aspects of the economic history of the United States. The subject areas ranged widely, from "population and immigration" to "the labor movement." Several divisions focused on large-scale commercial topics, including mining, manufactures, and agriculture and forestry. A division related to race and slavery was added in 1906.

The Institution published a number of useful works resulting from these efforts, most notably *History of Domestic and Foreign Commerce of the United States* (1915) by Emory Johnson and colleagues, *History of Manufactures in the United States* (1916–1928) by Victor S. Clark, and numerous indices of state documents. In addition, historic documents that otherwise might never have been saved were identified and, in the process, preserved for future scholars. On a more philosophical level, some people argue that the work of this Carnegie division provided a forceful rebuttal to Marxist economic histories that were appearing in increasing numbers at the time.

Nevertheless, the purview of the department was too sweeping to be reduced to the simplicity of "natural laws." In addition, the directors had little control over the sprawling confederation of investigators. It certainly didn't help that the researchers were unpaid. As early as 1911, these organizational problems were noted in the Institution's annual reports, and on December 11, 1916, despite President Woodward's dissent, the trustees voted to close the department. The department's director, Henry Farnum, retained hope that some of the more successful projects would be revived. "The past has many more lessons for us than has been hitherto realized," Farnum claimed, but his protests were in vain. The Institution had learned another lesson from the experiment; namely, that the physical and biological sciences were a better bet for funding.

DEPARTMENT OF TERRESTRIAL MAGNETISM
(1904–   )

Directors

LOUIS A. BAUER (1904–1929)

JOHN A. FLEMING, Acting Director (1929–1934); Director (1935–1946)

MERLE A. TUVE (1946–1967)

ELLIS T. BOLTON (1967–1974)

GEORGE WETHERILL (1974–1991)

LOUIS BROWN, Acting Director (1991–1992)

SEAN C. SOLOMON (1992–   )

The research vessel Maud *provides a backdrop for an observer making magnetic measurements in the Arctic Ocean above Siberia, 1922-25.*

The Department of Terrestrial Magnetism was formed in 1904 by Louis Agricola Bauer, a man of forceful convictions, to fulfill the goals set by Wilhelm Eduard Weber and Carl Friedrich Gauss of determining the earth's geomagnetic field and its temporal variations. Under Bauer's guidance, numerous land- and sea-based expeditions were made, including arduous journeys to the Arctic and remote regions of Australia, as well as seven ocean voyages by the nonmagnetic yacht *Carnegie.* The data accumulated during these journeys were extremely valuable— accurate enough to improve navigational charts (which included magnetic infor- mation) and complete enough to threaten the closing of the department during the 1920s on the grounds that it had achieved its mission.

Faced with closure, the department set about redefining its goals with a spirit as adventurous as that found among its early mariners and expedition leaders. Two young physicists, Gregory Breit and Merle Tuve, were already branching off in a new direction in 1925. Seeking to verify the existence of the ionosphere, they sent pulsed radio signals into the upper atmosphere and in the process discovered the principles of radar. Thereafter, new research directions were almost a way of life. During the 1930s and 1940s, the laboratory became a world-class center for the study of nuclear physics, which made fundamental discoveries about atomic forces. During World War II, scientists turned their attention to military goals and devel- oped the proximity fuse, a device that used radio waves to detonate antiaircraft artillery shells as they approached a moving target.

Odd Dahl (on ladder) and Merle Tuve working on the Van de Graaff genera- tor at the Department of Terrestrial Magnetism, 1935.

After the war, departmental scientists branched off into new fields, including isotope geology, seismology, astronomy, and biophysics. These diverse efforts resulted in discoveries regarding the structure of the Earth, age-dating techniques, the properties of genomes, and the existence of dark matter.

THE OBSERVATORIES OF THE CARNEGIE INSTITUTION OF WASHINGTON

(1904–   )

Directors

George E. Hale (1904–1923); Honorary Director (1923–1936)

Walter S. Adams (1924–1945)

Ira S. Bowen (1946–1964)

Horace W. Babcock (1964–1978)

Maarten Schmidt (1978–1980)

George W. Preston, Acting Director (1980–1981); Director (1981–1986)

Ray J. Weymann (1986–1988)

Leonard Searle, Acting Director (1988–1989); Director (1989–1996)

Augustus Oemler, Jr. (1996–   )

It is impossible to imagine the Carnegie Institution without the enthusiasms of George Ellery Hale. The moment he heard that Andrew Carnegie's institution of discovery was dispensing money, Hale decided he needed some. In 1904, he lobbied successfully for the establishment of a solar observatory on Mount Wilson in the San Gabriel Mountains outside Pasadena. The first telescope erected at this remote site was the Snow telescope, put into operation in 1905. It was here, in 1908, that Hale, aided by a spectroheliograph of his own design, discovered that sunspots are intense magnetic fields. This discovery was the first detection of a magnetic field beyond that of Earth.

Other telescopes followed: Two solar towers (the 60-foot erected in 1908 and the 150-foot completed in 1912) were supplemented by two magnificent stellar telescopes—the 60-inch, completed in 1909, and the 100-inch Hooker, which was tested successfully on November 2, 1917.

The Mount Wilson Observatory transformed the world of astronomy. It was with these magnificent instruments that Harlow Shapley mapped the globular cluster system of the galaxy and found its center. Edwin Hubble, widely recognized as one of the greatest astronomers since Galileo, captured the attention of the world with his discoveries that galaxies are distant, vast clusters of stars and that the universe is expanding. Walter Baade started a revolution that led to our understanding of the nature of stars, their life cycles, and the evolution of the

Milky Way galaxy. Walter Adams, Alfred Joy, Allan Sandage, and other Carnegie astronomers played a pivotal role by establishing the empirical basis of theories of stellar structure and evolution.

As current director Augustus Oemler points out, "Astronomers must be opportunists." They need the best possible telescopes to gather information about the distant universe. In 1948, a joint venture between Carnegie and the California Institute of Technology resulted in the dedication of the 200-inch telescope at Mount Palomar. Perpetuating the tradition of providing its scientists with excellent tools, the Institution decided to establish a new observatory in the Southern Hemisphere in the 1960s. Located in the Chilean Andes, the observatory was launched with 40-inch and 100-inch reflecting telescopes. Today, twin 6.5-meter reflectors are operated at Las Campanas Observatory by a Carnegie-led consortium of five American educational institutions: the Carnegie Institution of Washington, the University of Arizona, Harvard University, the Massachusetts Institute of Technology, and the University of Michigan. The telescopes' large fields of view and state-of-the-art instrumentation are enabling astronomers to study the chemical history of stars in our galaxy, as well as to learn more about the first galaxies to form near the edge of the universe. Carnegie scientists search for objects orbiting black holes, investigate galaxy collisions, and map the large-scale structure of the universe with these and other instruments around the world.

The Snow Telescope, 1912. Named for benefactor Helen Snow, this telescope was transported from Yerkes Observatory to Mount Wilson in 1905. The first photograph of a sunspot spectrum was taken with this telescope.

Following pages: 100-inch Hooker telescope at Mount Wilson, 1917.

GEOPHYSICAL LABORATORY

(1905–   )

Directors

ARTHUR L. DAY (1909–1918)

ROBERT B. SOSMAN, Acting Director (1918–1920)

ARTHUR L. DAY (1920–1936)

LEASON H. ADAMS, Acting Director (1936–1937); Director (1938–1952)

GEORGE W. MOREY, Acting Director (1952–1953)

PHILIP H. ABELSON (1953–1971)

HATTEN S. YODER, JR. (1971–1986)

CHARLES T. PREWITT (1986–1998)

WESLEY T. HUNTRESS, JR. (1998–   )

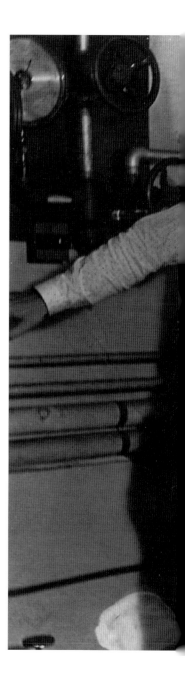

Since its inception in 1905, the Geophysical Laboratory has been devoted to fields of inquiry that transcend the boundaries of geology, physics, and chemistry. A state-of-the-art laboratory building, constructed with appropriately high ceilings and two-foot-thick walls, was built in 1907 on Upton Street in northwest Washington, D.C. There, over the next 80 years, Carnegie scientists made fundamental advances in understanding processes of the earth. In particular, they gained an improved understanding of the origins and distribution of rocks and minerals. Using X rays and an arsenal of spectroscopic techniques, they also studied the atomic structure of many natural compounds. When the moon rocks returned to Earth in 1969, Geophysical Laboratory researchers analyzed them. When University of Houston physicist Paul Chu needed to know the crystal structure of the first superconductor to work at liquid nitrogen temperatures, he sent the specimen to this laboratory for analysis. When the Freer Gallery of Art wanted to identify an ancient Chinese pigment, the laboratory was there.

Over the years, Geophysical Laboratory scientists developed many high-temperature and high-pressure technologies to assist them in their studies of the earth. Norman Bowen's experiments on igneous rocks paved the way for ever more sophisticated high-pressure apparatus, including the diamond-anvil cell that enabled Ho-Kwang Mao and Peter Bell to sustain 1,000,000 atmospheres pressure in a laboratory.

Geophysical Laboratory scientist Arthur Day and colleague (probably E. S. Shepherd) experiment using an electric arc furnace, ca. 1906. Before the Geophysical Laboratory's building was built, Carnegie scientists used facilities at the U.S. Geological Survey.

In 1953, Director Philip Abelson expanded the laboratory's research program by inaugurating investigations in biogeochemistry. Today, scientists have built on this tradition by initiating a research program in astrobiology, through which they seek to answer questions about the origin and distribution of life in the universe.

The Geophysical Laboratory moved to a new building on the Department of Terrestrial Magnetism's campus in 1990. At this new site, work continues in such areas as the development of new materials, the physics of high pressure, and the origin and evolution of life on Earth.

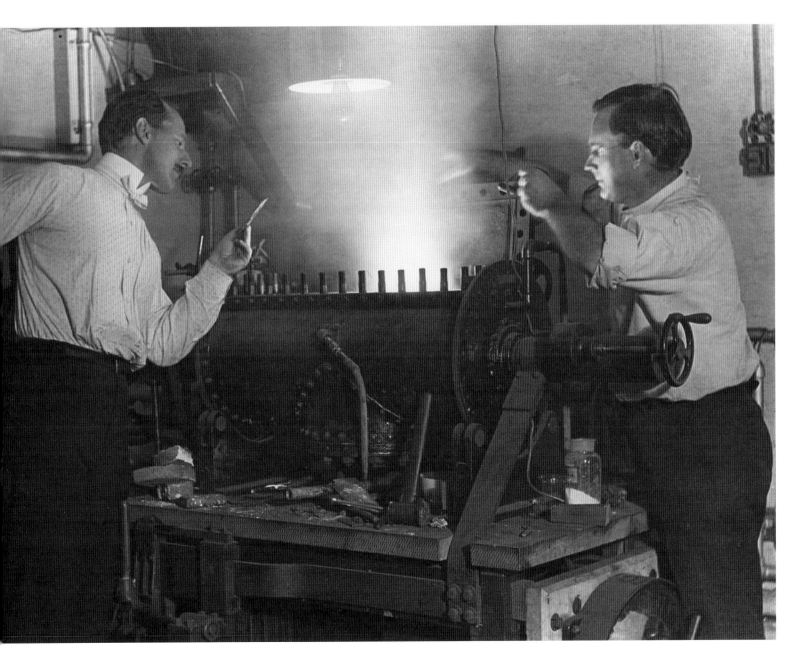

DEPARTMENT OF MERIDIAN ASTROMETRY
(1905–1938)

Directors

LEWIS BOSS (1907–1912)

BENJAMIN BOSS, Acting Director (1913); Director (1914–1938)

Dudley Observatory astronomer Heroy Jenkins with data sheets. Every hour at the telescope generated 20 hours of calculation.

In 1900, astronomers believed that if they could chart the stars overhead and measure their almost imperceptible motions relative to each other, they could begin to understand the structure of the universe. No one endorsed this philosophy more fervently than Lewis Boss, who served on the Carnegie Institution's astronomy advisory committee in 1902 and who, like other advisers, won funding for his own dream project.

Boss's idea was to create a state-of-the-art catalog of the positions and motions of stars as seen from Earth. While the catalog would draw on earlier observations, it would also include thousands of new observations.

This exercise was far from pedantic. If expertly collected, such data would transform human understanding of the heavens from a static, two-dimensional universe to the dynamic, three-dimensional one that is reality. Astronomers engaged in this work are called astrometers.

The Carnegie Department of Meridian Astrometry was authorized in December 1905. Based at Dudley Observatory in Albany, New York, the department's astronomers spent years observing, cataloging, and calculating. Then, in 1907, Boss launched a bold plan. He decided to dismantle the Albany telescope and move it to Argentina. There was good reason. No star catalog would be complete without the stars seen from the Southern Hemisphere.

With the help of Elihu Root, who was secretary of state and a Carnegie trustee, Boss received permission to build a temporary observatory in San Luis, Argentina. He surveyed the site during the summer of 1908. Then, after surviving a shipwreck on the return voyage, he oversaw the dismantling of the Olcott Meridian Circle telescope in Albany. The instrument was shipped, piece by piece, to Argentina. Meanwhile, an exact replica of the Albany observing-house had been built in San Luis—the same dimensions, the same sturdy piers, the same amenities (including no heat, so as not to interfere with the measurements). The plan was to make the data as consistent as possible.

The first observations at San Luis were made on April 6, 1909. Over the next 22 months, an astonishing 87,000 observations were made by astronomers working long shifts. It has been said that this work was a "mental sweatshop." The eye and the hand had to work tirelessly together to collect data. At least two members of the staff had to be sent home as a result of eye fatigue from the hectic pace.

Observing was only the beginning of the astrometer's work. For every measurement taken in the field, 20 hours of computation were required. The result of these efforts was the *San Luis Catalogue of 15333 Stars for the Epoch 1910,* published in 1928. Along with the *Preliminary General Catalogue of 6188 Stars for the Epoch 1900* (published in 1910) and the *General Catalogue* (published 1936–1937), this reference was an indispensable source of information on the Milky Way. Walter Baade called these works a "fantastic achievement," and Sir Arthur Eddington claimed that the measurements saved him an enormous amount of time in his own work.

But time was key, and time was running out on the field of astrometry. Long before the publication of the general catalog of stars, the world of astronomy was evolving. Astrophysics and cosmology, fields championed by George Ellery Hale and others, focused less on where celestial objects were located than on how they worked. Finally, in 1923, Edwin Hubble changed the world of stargazers forever. Astrometers thought they'd been studying the entire universe when, in fact, they'd simply been observing the suburbs of the sun.

Concluding that the Dudley astrometers had satisfactorily completed their work, the Institution closed the department in 1938. Nevertheless, the questions asked by Boss and his colleagues were important. In 2004, NASA plans to launch a satellite that will determine the positions of more than 40 million stars.

Olcott Meridian Circle Telescope installed at San Luis, Argentina, 1908.

NUTRITION LABORATORY

(1907–1946)

Directors

FRANCIS G. BENEDICT (1907–1937)

THORNE M. CARPENTER (1938–1946)

The Carnegie Institution began making grants in nutrition in 1903. Francis Benedict, a professor at Wesleyan University, received his first grant in 1905. A year later, President Robert Woodward concluded that Benedict's work was significant enough to receive large-scale funding, and in 1907 the Nutrition Laboratory was organized with Francis Benedict as director. After investigating a number of sites for the new venture, Benedict and his colleagues chose land near the Harvard Medical School in Boston. The building was completed in 1908. At that time, the laboratory received heat, light, and refrigeration from the power plant of the Harvard Medical School.

From the outset, the mission of the laboratory was "to conduct fundamental scientific investigations in vital activity with special reference to the laws governing total metabolism, heat production, heat elimination, and heat regulation." In other words, investigators were less interested in diets than in how animals processed food to produce heat for body processes and muscular activity. The term "animals" was interpreted in a broad sense. Human beings were studied intensively. In addition to establishing standards of basal metabolism for "normal humans" of both sexes and of varying ages, the laboratory studied the effects of extreme conditions on human biological systems. These conditions included the effects of fasting, athletic activity, and fatigue. In 1913, a study of the physiological and psychological effects of ethyl alcohol on humans began. This landmark work continued for many years, even after the advent of prohibition in 1920.

Scientists at the Nutrition Laboratory realized that data for humans could be supplemented by data for other animals, and they expanded their research program accordingly. Rabbits, cattle, rats, snakes, tortoises, woodchucks, and monkeys are a few of the organisms that were studied by laboratory scientists and their colleagues over the years. In one extraordinary project, an investigation of the metabolism of a four-ton elephant was supplemented by studies of the physiology of groups of elephants.

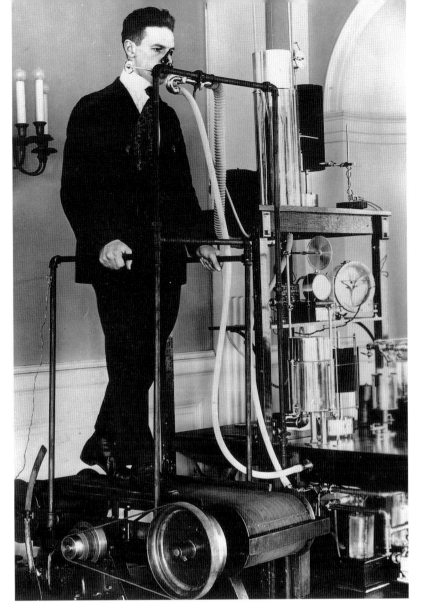

Man with breathing apparatus at the Nutrition Laboratory, Boston.

These investigations required sophisticated devices to measure the variables under study; namely, heat production, heat elimination, respiratory exchange, and surface and internal body temperature. The laboratory staff designed, built, and tested a range of respiration calorimeters, respirometers, and thermometers. Much of this work was carried out in the laboratory's own machine shop.

Throughout the Nutrition Laboratory's existence, its scientists maintained active collaborations with other institutions, including the Carnegie Institution's Departments of Embryology and Genetics. In fact, cooperation was a hallmark of Benedict's research philosophy. Many visiting investigators spent time at the Boston facility.

The work of this department was broad—too broad for a single institution with limited means. With Benedict's retirement in 1937, the laboratory lost its best advocate. During World War II, the laboratory was almost exclusively involved in contract work for the government. Then, in 1946, the department was closed; its work, in the words of the Institution's president Vannevar Bush, was "ably supported by other agencies."

DEPARTMENT OF EMBRYOLOGY

(1914–   )

Directors

FRANKLIN P. MALL (1914–1917)

GEORGE L. STREETER, Acting Director (1918); Director (1919–1940)

GEORGE W. CORNER (1941–1955)

JAMES D. EBERT (1956–1976)

DONALD D. BROWN (1976–1994)

ALLAN C. SPRADLING (1994–   )

The Carnegie Institution's Department of Embryology, located on the Johns Hopkins University campus in Baltimore, was founded in 1914 to study biological development in humans. Originally, much of the research centered on a remarkable collection of human embryos amassed by director Franklin Mall and his successors. The collection ultimately became so complete and instructive that

it provided the "bureau of standards" for embryological research worldwide. Elizabeth Ramsey's landmark work on the placenta was done at this department, as was George Corner's work on tissue culture. Under Corner's leadership, the department established and maintained one of the world's first monkey colonies, an extraordinary resource that allowed Carl Hartman to elucidate the monkey menstrual cycle.

By the middle of the twentieth century, classical embryology began to give way to research based on the new techniques of molecular analysis. James D. Ebert, who assumed the directorship in 1956, attracted a number of creative young researchers. Using a variety of organisms, these scientists continued to tackle the central question of embryology; namely, how does a multicelled organism arise from a single egg? Donald Brown, for instance, was one of the first scientists to isolate an animal gene. The work of Allan Spradling and Gerald Rubin on gene transfer in *Drosophila*—the first instance of gene transfer in a higher organism—has led to better understanding of mammalian development, including that of humans. Before joining the department in 1983, Joseph G. Gall discovered gene amplification (a discovery made simultaneously by Donald Brown and Igor Dawid). He went on to study gene transcription in newt and frog oocytes.

Today, the Department of Embryology is relatively small, with just eight staff members. The scope of its research, however, is broad. Departmental scientists study a range of organisms, including yeast, fruit flies, worms, zebra fish, mice, and frogs. The novelty and breadth of the research at the department is enhanced by a program of staff associates, which allows young researchers to head their own laboratories for five years.

Modeling room at the Department of Embryology, 1921.

that in the truest sense he is the uncommitted investigator, suitably endowed and suitably protected, whose time, quite literally, is bought by the Institution and then returned as unconstrained endowment."

Both versions of the concept sound good in theory, but surprisingly the rhetoric has been transformed fairly successfully into policy. If one were to depict the Carnegie Institution as a pyramid, there is no question that the scientists would occupy the top section and the financial and support staffs would occupy the middle and base sections. The focus on the individual investigator has been institutional policy since the beginning, but by the second half of the twentieth century the wisdom of perpetuating this approach was obvious. With a minimum of bureaucracy and a streamlined system of communication, scientific staff would be free to get on with their work without distraction. Increased productivity was all but inevitable.

But that's not all. There is a less obvious but possibly more important advantage to this mode of support. An organization that emphasizes the work of individuals (as opposed to the discipline-based groups of scholars that define most university departments) is more likely to embark on exciting new fields of inquiry in an expeditious manner. As Haskins put it, Carnegie scientists enjoy "freedom from fixed commitment." This freedom, in turn, not only facilitates the initiation of new ventures, but it also encourages cross-disciplinary studies that are often crucial for scientific advances. Without such freedom to experiment and explore, for instance, it is impossible to conceive how a group of nuclear physicists at the Department of Terrestrial Magnetism could have shifted gears in the 1950s and embarked on their landmark study of the bacterium *Escherichia coli.*

For this system to work well, of course, a core of gifted and discriminating investigators is required. Since their scientific questions essentially drive the work of an entire institution, they'd better be asking good questions! Two additional policies help ensure that top-notch scientists are employed. First, Carnegie staff positions are few. Two departments, Embryology and Plant Biology, have just eight full-time staff members and the larger departments support about 15. Intense interviewing procedures and scrupulous oversight by departmental directors help ensure that recruits will be world-class. In addition, there is no formal tenure system at Carnegie. Each staff scientist undergoes a rigorous peer review every five years.

Embrace Choices and Change

Not once in 100 years has anybody at the Carnegie Institution lamented, "We have more money than we know what to do with!"

Tough choices about whom to fund were made from the outset. Not surprisingly, however, choices were more necessary than ever in the postwar years. No one in a position of authority seems to have advocated the notion that, to maximize leadership in a particular field, the Institution should close all departments but one. In other words, a commitment to a number of fields was a given. The question was, how many researchers and which fields of inquiry?

A precedent to terminate ventures that were essentially "complete" or that could be better carried on by others was well in place by the end of World War II. On the latter basis, the Nutrition Laboratory was closed in 1946. Twelve years later, in 1958, the Department of Archaeology, the last remaining section of the Department of Historical Research, was terminated for the same reason. The Department of Genetics survived a little longer, but in 1971 it too was closed. The justifications for this latter decision are various and complex, but choices, it would seem, are choices, whatever the results. Vannevar Bush preferred the word "taperings" for his own terminations, and in a sense, it's a valid distinction. Despite formal closings of the Departments of Archaeology and Genetics, several staff members continued to receive institutional support until they retired. They continued to produce good work.

Closing departments to save money and concentrate institutional resources on the most productive research programs had its benefits, of course. But to remain vibrant and forward-looking, an institution also has to consider starting new ventures, possibly even expensive ventures. Into this category falls the Institution's decision to build an optical observatory in the Southern Hemisphere.

Carnegie astronomers were the driving force behind this decision. As they pointed out, there was no such facility, and they needed data that could be obtained only from the southern skies. For instance, they wanted to see the center of the Milky Way, an impossibility from ground-based telescopes in the Northern Hemisphere. To answer fundamental questions about the formation of galaxies and the structure and dimensions of the universe, Carnegie astronomers needed the best equipment in the world.

The Institution could have stepped aside and allowed others to build such a facility. However, if it wanted to maintain its leadership in the field of astronomy, it had no choice but to embrace George Ellery Hale's goading rhetoric: "make

Excavations at Mayapán, the last project undertaken by Carnegie archeologists, had many innovative features, including the mapping of the walled urban center. The drawing of the temple reconstruction is by Tatiana Proskouriakoff. The Mayapán report was published in 1962, several years after the closing of the department.

Crawford Greenewalt (front left) and Philip Abelson at the dedication of the duPont telescope, October 6, 1976.

The duPont telescope at Las Campanas, Chile, 1976.

no small plans." Caryl Haskins endorsed the Hale philosophy in this context, as did the board of trustees. Accordingly, the Carnegie Southern Observatory (CARSO) was launched in 1969.

The Institution chose a site in Chile, about 80 miles northeast of the coastal town of La Serena. Twenty thousand hectares were acquired from the Chilean government at a place called Las Campanas, and by 1970 a 20-mile road had been completed. A 40-inch telescope, funded by a gift from astronomer Henrietta Swope, was placed in operation in 1972. Four years later, the 2.5-meter duPont telescope was dedicated. Funded by Carnegie trustee Crawford Greenewalt and his wife, Margaretta, this telescope was designed by Carnegie astronomers.

But the CARSO project didn't stop there. In 1993, the Institution began construction of twin 6.5-meter telescopes, designed to be connected by a 60-meter underground tunnel that would permit beam combining and interferometry. With such equipment, astronomers would have some of the "best seeing" in the world. Named the Magellan Project, these twin telescopes called for collaborations with scientists from other institutions, including Harvard University, the Massachusetts Institute of Technology, and the Universities of Arizona and Michigan.

Big budget? Yes. Big questions? Absolutely. But the assumptions underlying the Las Campanas program were typical of the small science Carnegie had always supported—namely, work defined by creative investigators whose primary purpose was to add to the world's stock of knowledge.

Encourage Pursuit of "Outside" Grants but Not Too Many

By the mid-1970s, the Institution's endowment was in trouble. Its value had dropped dramatically, largely because of the decline in the stock market, changes in financial management strategy, and inflation. The investment in Las Campanas didn't help financial matters either. Now, if ever, there was good reason to consider the question of federal support for Carnegie science.

Most institutions had followed this route years earlier, but not the Carnegie Institution, whose scientists valued independence above almost everything. Therefore, President Philip Abelson approached the question cautiously. Abelson, who had moved through the ranks of the Institution from scientist at the Department of Terrestrial Magnetism, to Geophysical Laboratory director, to Institution president, understood his colleagues' reluctance in pursuing outside money only too well. And he agreed that there was much to worry about: the erosion of the unique character and flexibility of the Institution; the temptation to undertake fashionable work that would be readily funded rather than innova-tive, long-term efforts of the individual researcher; the diversion of valuable research time to the tedious application process; and the weakened attachment of a staff that had come to rely on institutional support and now might perceive themselves to be sold out.

The Department of Terrestrial Magnetism and the Geophysical Laboratory at the Broad Branch Road campus, 1991.

Nevertheless, reporting to the trustees in January 1977, Abelson proposed "to move cautiously using more federal funds to replace Carnegie funds." To protect the independence the Institution's scientists, he imposed two requirements on this action: Staff salaries must not be paid by grants, and no staff member should be required to get outside grants. Thus the Institution edged slowly but surely into the world of big-science funding. Over the next 20 years, the trend would expand, though it would always be controlled by administrative policy that carefully monitored external funding. Today, some staff salaries are partially funded by grants, and almost all staff scientists submit grant proposals regularly. In the recent past, approximately one-third of the annual operating budget has come from federal grants, but there is no hard-and-fast rule. In these ways, the Institution hopes to augment its resources without sacrificing its independence.

Grab Your Partner

One of the most powerful ways to leverage the impact of any small organization is to form partnerships with other organizations with similar goals. Provided the partnerships preserve the integrity of the individual groups, such collaborative arrangements offer attractive opportunities to pursue projects that are too large and ambitious for any single organization.

At the Carnegie Institution, this approach has been endorsed by researchers across the board. For instance, as mentioned above, the Magellan Project is a col-laborative venture by astronomers based at five institutions. Similarly, for nearly 20 years, a collaboration of scientists at Princeton University, the State University of New York at Stony Brook, and the Geophysical Laboratory has studied effects of high pressure on a variety of materials under the organizing title, "Center for High-Pressure Research." The sequencing of the genomes of the model plant *Arabidopsis* and model animal *Drosophila*—remarkable achievements—were the work of international teams that included members of the Departments of Plant Biology and Embryology, respectively. In another trailblazing venture, Carnegie scientists are members of a diverse group of organizations sponsored by the National Aeronautics and Space Administration that focus on studies of the origin

and distribution of life in the universe. As befits such a huge question, there are a dozen institutional "nodes" to this effort, and each node (of which the Carnegie Institution is one) includes additional collaborators.

Even within the Carnegie Institution itself, partnering has been part of the program. In the early 1980s, President James Ebert raised a provocative question: Why would a small group of scientists doing good work at one site fail to do even better work if it joined forces with a similar group elsewhere?

The outcome of this suggestion was the co-location of the two Washington-based Carnegie laboratories investigating the physical sciences, the Geophysical Laboratory and the Department of Terrestrial Magnetism, on the latter's campus. Construction of new laboratory facilities began shortly after the board of trustees endorsed the co-location in 1988, and in 1990 the Geophysical Laboratory, under the leadership of Charles Prewitt, moved into its new home. As Ebert predicted, many fruitful collaborations among Carnegie scientists and their students ensued.

With constant improvements in electronic communications systems, it is likely that cooperative ventures of many sorts will increase in the future at Carnegie.

Let There Be Light

One of the early goals of the Institution was to support students who came to Washington to study. This program enjoyed limited success, but it was succeeded by a very effective educational initiative that has augmented the impact of individual Carnegie scientists far beyond their own laboratory walls. This initiative is the postdoctoral fellowship program. Inaugurated in 1944, this productive program encourages young investigators to study with established scientists in their fields before moving on to begin scientific careers of their own. The Department of Embryology has elaborated on the basic concept and instituted an innovative program of staff associates, which provides gifted young researchers at the beginning of their careers their own laboratories and independence for five years.

That Carnegie scientists benefit intellectually from these interactions is undeniable, but it is also true that the Institution benefits in terms of broadened exposure and impact. As Carnegie postdocs move on to permanent positions, they acquire students of their own, and the influence of Carnegie staff members expands. Ironically, the founders of the Institution wanted to protect researchers from teaching duties, but the postdoctoral fellowship program, which takes a broad view of teaching, has paid enormous dividends for Carnegie science.

Students attending First Light program at the Carnegie administration building, 1996.

Following page:
Hunting the magnetic pole, an observer (probably L. A. Bauer) takes measurements of unusual magnetic disturbances at Treadwell Point, Alaska, 1907.

This outcome would most likely have pleased Andrew Carnegie enormously. Carnegie was as true a believer in the power of education as one is likely to find this side of the nineteenth century. Contrary to myth, Carnegie never stipulated that his portrait be hung inside the library buildings he built. What he did express was his hope that the builders of the libraries would place an image of the sun with the accompanying motto, "Let there be light," at the entrance of each library, an effective reminder of the importance of learning.

Maxine Singer, who became president of the Institution in 1988, has built on his philosophy in a creative and effective way. Shortly after she arrived, Singer began to plan a Saturday science school for elementary-school children called First Light. Housed in the administration building, this program introduces third- to sixth-graders to the world of science and its ways of knowing. Beginning in 1993, Singer expanded this effort to include training for science teachers through the establishment of the Carnegie Academy for Science Education, or CASE.

Thus the Institution enters its second century as it began its first: seeking and disseminating knowledge for the benefit of humankind.

PART TWO

BUILDING
THE EDIFICE OF
KNOWLEDGE

*I*n 1945, Vannevar Bush published an essay that compared scientists to builders. Some workers, Bush wrote, pile up "odd blocks" without caring whether they fit anywhere. Others spend their time arguing about the exact arrangement of a cornice or abutment. Still others, Bush said, "spend all their days trying to pull down a block or two that a rival has put in place."

And then there are the people of vision. These are the builders who "can grasp well in advance just the block that is needed for rapid advance on a section of the edifice to be possible, who can tell by some subtle sense where it will be found, and who have an uncanny skill in cleaning away dross and bringing it surely into the light." These people Bush dubbed the "master workmen."

The Carnegie Institution has supported many master workers over the years. In the following chapters, the reader will meet some of these exceptional men and women and, in the process, sample some of the best work produced by Carnegie scientists during the last century. These investigators have worked in diverse fields of inquiry—including astronomy, earth science, biology, and such cross-disciplinary studies as biophysics. All, however, have been curious about the workings of the natural world and have had the creativity, equipment, and vision to begin to find answers.

Their "good seeing" has changed the way the rest of us see the world.

J. S. Potter and Margaret D. Findley studying mouse leukemia at the Department of Genetics, ca. 1935.

ASTRONOMY

BUILDING TELESCOPES

*S*cientific understanding grows in many ways. New insight may come from flashes of theoretical vision or the slow accumulation of routine data, from mathematical manipulations or meticulous measurements. But, as often as not, great scientific discoveries are facilitated by the design and construction of great instruments. Such was the legacy of astronomer George Ellery Hale.

George Ellery Hale was an interesting mix—part eighteenth-century gentleman scientist, part twentieth-century entrepreneur. His father, William Ellery Hale, manufactured hydraulic elevators in Chicago at a time when the city was building skyscrapers at a phenomenal rate. This private fortune not only allowed Hale to be free of worry about employment, but it also gave him resources to call on when he needed new (and expensive) instruments for his astronomical work. Following in the long tradition of wealthy men like William Halley and Henry Cavendish, he used his private fortune to advance the cause of science.

On the other hand, Hale was also a consummate fund-raiser and organizer. He was a first-rate scientist, one of the great pioneers of solar astronomy, but he is remembered today chiefly for the institutions that he founded and nurtured: the Mount Wilson and Palomar observatories, the National Research Council, and the California Institute of Technology, to name a few of the most important.

Andrew Carnegie and George Ellery Hale at the 60-inch telescope at Mount Wilson, 1910.

In this regard, Hale was the precursor of a type of scientist who would become increasingly important during the twentieth century: the organizer and entrepreneur who not only could do good work but could marshal and organize the resources needed to do it. In a time when even a mundane experiment in some areas of science can cost many millions of dollars and involve dozens (if not hundreds) of people, these talents have become an essential part of research.

Hale's family lived in the Hyde Park section of Chicago, then a wealthy suburb along Lake Michigan and later home to the University of Chicago. From an early age he was interested in the tools of astronomy, and encouraged by his father, he attempted to build a telescope. This early effort was unsuccessful, but so ardent was the young Hale about needing a telescope that his father eventually purchased a 4-inch Clark telescope, then a state-of-the-art piece of equipment, for use in the family's backyard observatory.

As an undergraduate at the Massachusetts Institute of Technology, Hale worked as a volunteer in the Harvard College Observatory. In July 1889, while riding on a Chicago trolley car, an idea came to him "out of the blue" about an instrument that would allow him to photograph prominences on the sun in full daylight. The construction of such an instrument had been a goal of astronomers for a quarter-century, and in his MIT senior thesis Hale demonstrated that his design for this instrument, called a spectroheliograph, was feasible. This idea marked the beginning of a research career that later caused astrophysicist Robert Howard to remark that "Hale may be said to be the father of modern solar observational astronomy."

After his marriage (and a honeymoon visit to Lick Observatory in California), Hale settled in Chicago. In a portent to things to come, he persuaded his father to buy a 12-inch telescope that became the centerpiece of a small private observatory behind the family home. With this instrument, Hale produced the first photographs of some of the detailed features of the sun's surface. Hale also became an associate professor at the newly founded University of Chicago (although he never got around to finishing his Ph.D.).

It was during this period that he made the acquaintance of Charles Yerkes, a Chicago streetcar magnate, and convinced him to fund a new observatory that would house the world's largest telescope at the time. When the Yerkes Observatory was dedicated at Williams Bay, Wisconsin, in 1897, it became a major center for international astronomy research. The 40-inch Clark telescope, a beautiful as well as useful instrument, was then the world's largest refracting

George Ellery Hale operating his spectroheliograph, which he used to study magnetic fields in sunspots.

telescope (that is, a telescope in which light is focused by a large glass lens) and remains so today. In 1894, Hale oversaw the founding of the *Astrophysical Journal*, still the premier professional journal in astronomy, and in 1899 the first meeting of the organization that would become the American Astronomical Society was held at Yerkes.

Astronomy in the Last Quarter of the Nineteenth Century

With someone like Hale, whose impact on the institutions of twentieth-century science was so profound, it is easy to forget the contributions he made as a scientist. When he graduated from MIT in the early 1890s, the field of astronomy was beginning to undergo a profound transition. Simply stated, it was changing from astronomy (where are celestial objects and how are they to be classified?) to astrophysics (how do those objects work?). As physics and chemistry developed,

and as astronomers became able to identify chemical elements in distant stars, a whole new class of questions could be asked. Some of the answers to these questions would come from looking at the sky through a telescope, of course, but others would come from experiments in earthly laboratories. These latter studies, linking atoms in laboratories to atoms in stars, provided the core of the new discipline of astrophysics.

Throughout his career, Hale was a tireless advocate of the science of astrophysics. It was no accident, for example, that his new publication was called *Astrophysical Journal* and not the *Astronomical Journal*. He argued that the best way to understand the stars was to examine the one closest to us, our own sun. His early work on Mount Wilson concerned the nature of sunspots. His team established that sunspots were regions of lower temperature than their surroundings. By careful comparison of light emitted by atoms in sunspots to light emitted by those same atoms held between the poles of a powerful magnet in Pasadena, Hale was able to establish that sunspots are regions of intense magnetic fields. This was, in fact, the first detection of a magnetic field somewhere other than the

Mount Wilson Observatory, 1931. Included in this view, left to right: the horizontal Snow telescope, the 60-foot solar tower, the 150-foot solar tower, the 60-inch telescope dome, the 50-foot interferometer building, and the 100-inch telescope dome.

earth. Much of his research for the rest of his career involved understanding the nature of the sun's magnetic field (although the definitive measurements of the field strength weren't made until the early 1950s).

"Make No Small Plans"

In 1894, Hale talked his father into funding a 60-inch mirror for a reflecting telescope (that is, a telescope is which light is focused by a mirror rather than a lens). For several years he was unsuccessful in finding the additional funding needed to build a new observatory, but as an adviser to the newly founded Carnegie Institution of Washington in 1902, he was able to make his case. In 1904, after a long and complex campaign that at one point required Hale to put up $30,000 of his own money, he was successful and, on December 20, 1904, he witnessed the founding of the Mount Wilson Solar Observatory on a mountaintop near Pasadena, California. Arguably, this was the most important installation with which either Hale or the Carnegie Institution was ever associated, since it was here, with the 100-inch Hooker telescope, that Edwin Hubble later mapped the cosmos.

Even before his 60-inch telescope was completed in 1908, Hale had installed equipment borrowed from Yerkes on Mount Wilson and made important discoveries about the sun. Then, in 1904, he met Los Angeles businessman John Hooker and an even bigger plan emerged. Hale proposed a still larger telescope—one with a mirror a full 100 inches across. Hooker, who had made a fortune in the hardware and steel pipe business, agreed to pay for the mirror. But that was only part of the project. Hale had to go through another complex set of negotiations with the Carnegie Institution to obtain funding to complete the telescope. In addition, there were problems with casting the large block of glass needed for the mirror: The first blank, when it arrived in Pasadena, was deemed unusable because of the presence of bubbles, and the second cracked while cooling at the foundry in France. At one point, Hale's wife exclaimed in exasperation, "I wish that glass was in the bottom of the ocean!" Eventually, however, the first blank was found to be usable, bubbles and all, and, in 1911, Andrew Carnegie made an additional gift to his Institution with the strong suggestion that it be used to support Hale's project. "I hope," he wrote using simplified spelling, "the work at Mount Wilson will be vigorously pusht, becaus I am so anxious to hear the expected results from it. I should like to be satisfied before I depart, that we are going to repay to the old land some part of the det we owe them by revealing more clearly than ever to them the new heavens." The telescope went into operation in November 1917.

Building Institutions

It would be wrong to suppose that during this period of extraordinary progress Hale's energies were focused solely on astronomy. He became a trustee of the Throop Polytechnic Institute, a trade school in Pasadena. With the help of James Scherer, whom Hale recruited as the Institute's new president, he built Throop into a world-class technical university, known today as the California Institute of Technology. In 1915, with U.S. entry into World War I imminent, Hale offered the services of the National Academy of Sciences to Woodrow Wilson. This initiative led to the founding of the National Research Council, which remains the research arm of the academy. He also was instrumental in the construction of the beautiful headquarters building for the academy, which was completed in 1924. Located on the Washington Mall across from the Lincoln Memorial, it was funded as part of a $5 million grant to the research council from the Carnegie Corporation.

Hale also had a dream about transforming Pasadena into a major cultural center, and he urged his friend Henry Huntington to use his fortune and his art collection to establish a center for the study of the humanities. In 1919 Hale was made a trustee of what would become the Huntington Library and Art Gallery, which Huntington endowed in 1927 and which remains a major scholarly and artistic center today.

In 1923, Hale's health, always fragile, forced him to resign from the directorship of Mount Wilson. In his "retirement" he founded the Hale Solar Laboratory in Pasadena and successfully approached the board of the Rockefeller Foundation with his plan for a 200-inch telescope to be built on Mount Palomar in southern California. The foundation donated $6 million for the telescope to be built and run jointly by the Carnegie Institution of Washington and the California Institute of Technology, a partnership that lasted until 1980.

Hale died before that telescope, delayed by World War II, was dedicated in 1948 and named in his honor. Astronomy in the last half of the twentieth century was dominated by the telescope named for the man who created so many of the century's leading institutions.

THE FABRIC OF THE UNIVERSE

Cosmology, the study of the origin and evolution of the universe, lies at the heart of the human quest for understanding. Yet, until the 1920s, cosmology was more a subject for speculative philosophy than observational science. Prior to the dedication of the 100-inch telescope at Mount Wilson, astronomers had no way to resolve the most distant objects in the heavens. They had no yardsticks with which to measure the scale of creation, nor could they shed any light on its distant past.

Stargazing in the Early Twentieth Century

In the first decades of the twentieth century, the transition from classical astronomy to astrophysics was well under way. Aided by improved telescopes, astronomers had measured the distance to the stars and established the vast emptiness of our immediate neighborhood in the universe. But there remained the problem of the nebulae.

"Nebula" is the Latin word for "cloud," and it aptly describes many objects seen in telescopes, objects that appear to be foggy, smeared-out patches of light. The nebula puzzle had been debated by astronomers for decades. Were these

fuzzy patches of light relatively close to Earth, or were they far away and only appeared to be smeared out because the available telescopes could not resolve their features? Just as fine print held far away ceases to look like words and letters, a celestial object might appear featureless even though it actually had a definite structure. This was not just an idle question because it dealt with the basic structure of the universe. Was all matter gathered into the single large complex we call the Milky Way? If so, then all of the nebulae must be part of that complex. Alternatively, to use the felicitous phrase coined by Immanuel Kant, is the Milky Way just one of many "island universes?" If so, then some of the nebulae could be those other "universes" (or galaxies, to use the modern term) and their fuzziness would simply be due to their great distance from Earth. By Hubble's time, astronomers had gathered enough evidence to persuade them that at least some nebulae were in the Milky Way. But there remained a large number whose identity was unresolved. They appeared as faint, roundish patches of light in the telescope field, and it was on these objects that the debate centered.

As often happens in the sciences, a question of great philosophical import came down to a grubby question of experimental detail: How can we measure distances to nebulae? If they were all less than about 80,000 light-years away (the estimate at the time for the size of the Milky Way), there was only one galaxy in the universe. If, on the other hand, they were farther away than the outer edges of the Milky Way, the nebulae must be other galaxies, and the universe would be far vaster than anyone had imagined. The challenge was to measure the distances to astronomical objects.

This was a far from trivial problem. The sky presents itself to us as a two-dimensional array, and there is no way of telling how far away something is just by looking at it. An object might appear faint, for example, because it actually emits only a small amount of light. However, it could equally well be emitting huge amounts of light but be very far away.

The easiest measurements involve planets and stars relatively close to Earth. For these objects, astronomers use the old-fashioned method of triangulation. By watching the apparent motion of a star against the distant background as Earth moves in its orbit around the Sun and using a little basic geometry, the distance to the object can be measured directly. The problem with this method is that, as the distance to the object gets very large compared to the diameter of Earth's orbit, the apparent movement of the star gets smaller and smaller until, eventually, it is lost in the fuzz imposed by the limits of the telescope. Today, even with high-power telescopes and satellites, triangulation can only be used to establish distances to about 600 light-years—a mere hop, skip, and jump on the cosmic scale.

Beyond triangulation, other, less direct methods of distance determination must be used. The most important of these depends on a variety of objects called "standard candles." The total energy radiated by these objects is known by one means or another. A 100-watt lightbulb, for example, is a splendid standard candle. If we see such a bulb far away, by comparing the amount of light we receive with the known amount radiated, we can deduce the bulb's distance. This calculation involves working out how far away the object would have to be for the light to be diluted by the measured amount. The astronomical distance problem, then, comes down to finding the equivalent of a 100-watt lightbulb in the sky.

Henrietta Leavitt (1868–1921) of Harvard College Observatory had recognized the existence of just such a class of objects called Cepheid variables, a class of stars that go through a steady brightening–dimming–brightening cycle over a period of days to months. Their name comes from the fact that the first one to be studied was in the constellation Cepheus in the Southern Hemisphere. Leavitt established that the period of the cycle is directly related to the total emitted energy, or luminosity, of the star—the longer the period, the greater the star's intrinsic brightness. Leavitt's discovery led to the development of a simple galactic yardstick based on the observed brightnesses and periods of Cepheid variables. By monitoring the period of the cycle, astronomers could determine the star's actual luminosity and then calculate its distance from Earth. If the Cepheid variable is in a larger structure (a star cluster, for example), this distance is also a reasonable estimate of the distance to everything else in the structure. It was the use of this distance scale that allowed astronomers to measure the size of the Milky Way. However, until the 100-inch Hooker telescope began to collect data at Mount Wilson, individual Cepheid variable stars in nebulae couldn't be distinguished, even if they were actually there. Thus there was no way of knowing how far away the nebulae were.

Edwin Hubble (1889–1953)

Edwin Hubble led an unusually adventurous life for a man who was destined to become one of the most famous astronomers of the twentieth century. Born in Missouri, he grew up in the town of Wheaton, Illinois, then a railroad suburb of Chicago. As an undergraduate at the University of Chicago, he excelled at both

Edwin Hubble and Sir James Hopwood Jeans at the 100-inch telescope.

academics and athletics. He played on the university's championship basketball team in 1908. He was also a skilled boxer, and he claimed to have been approached by promoters who wanted him to turn professional and take on the reigning heavyweight champion, Jack Johnson. He reportedly appeared in an exhibition match with the French heavyweight champion Georges Carpentier.

To supplement his college scholarship, Hubble served as a laboratory assistant for Nobel Laureate Robert Millikan and worked on railroad surveying crews in the rugged North Woods during the summers. After receiving his B.S. in 1910, he went to Oxford on a Rhodes scholarship. There he studied Roman and English law, and when he returned to the United States in 1913, he accepted a teaching position at a girls' high school, where the young students were enraptured by their new and dashing Spanish teacher. After teaching for a year, however, Hubble's abiding interest in astronomy surfaced. His father, who had been unimpressed by his son's starry interests, had died by this time. Hubble returned to Chicago for graduate work.

George Ellery Hale, who was an occasional visitor at Yerkes Observatory, where Hubble carried out his observational work, was sufficiently impressed to invite the young man to Mount Wilson. But Hubble decided to join the army and postponed joining the Carnegie staff until 1919, when World War I was over and he had returned from Europe. When he arrived at Mount Wilson he went right to work, using the 100-inch telescope to study the Cepheid variable stars. Over the next two decades, the collaboration among Hubble, his assistant Milton Humason, and the Hooker telescope transformed our understanding of the universe.

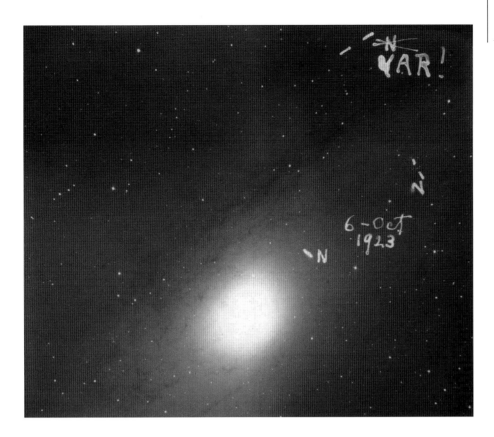

Left: Galaxies photographed by Hubble at Carnegie's 100-inch telescope: NGC 1097 and NGC 5236.

Right: On the night of October 5-6, 1923, Hubble photographed the Andromeda Nebula and on the photographic plate marked three objects in that nebula with the letter "N." At first, he believed they were stars called "nova," but he soon realized that one was a Cepheid variable star that could be used to measure the distance to Andromeda. Look for the star that Hubble relabeled "VAR!"

Hubble's pioneering work led to three major discoveries, any one of which would have secured his reputation. One of these contributions was the development in 1926 of a classification scheme for galaxies that, as amplified by Carnegie astronomer Allan Sandage, is still in use today. Hubble observed that galaxies come in many shapes. Some, like the Milky Way, are swirling spirals of stars. Others look like huge elliptical footballs, while still others are small and irregularly shaped. Hubble proposed a kind of zoology of the cosmos to facilitate the cataloging of the diverse objects he observed in his eyepiece. Classification schemes may not seem very exciting, but they usually represent important frameworks that yield original insights around which many other scientists can organize their thinking.

In another fundamental contribution, completed in 1924, Hubble established once and for all that the universe is made up of a collection of galaxies. Using the unparalleled power of the 100-inch telescope, Hubble distinguished individual Cepheid variable stars in a galaxy known as Andromeda or M31 (the designation comes from the fact that it is the 31st object in a catalog compiled in the eighteenth century by the French astronomer Charles Messier). Using these variable stars as standard candles, Hubble began to establish the incredible distances that separate the galaxies. His value for the distance to M31, for example, was 490,000 light-years, considerably larger than the diameter of the Milky Way. (This initial estimate was, in fact, too small; the current accepted value for the distance to M31 is 2.2 million light-years.) Hubble's results confirmed the notion that some of the nebulae were far beyond our own Milky Way and, in the process, greatly expanded human understanding of the size of the known universe.

Hubble's discoveries of nebular characteristics—their varied shapes and immense distances—were fundamental contributions, firmly rooted in the tradition of centuries of painstaking astronomical observations. But his most remarkable finding transcended the mere description of astronomical objects. Hubble discovered the astonishing fact that the universe is expanding. By presenting compelling observational evidence for the age and origin of the universe, he transformed cosmology from philosophical speculation to hard science.

Hubble's great insight came when he combined his own measurements of distances to nearby galaxies with measurements of the light coming from those galaxies that had been made by other astronomers. Others had observed that the light emitted by atoms in those galaxies seemed to be of longer wavelength (i.e., redder) than light emitted by those same atoms in laboratories on Earth. This so-called "red shift" is easily explained as an example of the Doppler effect, the same effect that makes the sound of a car engine change as it passes you. According to this interpretation, light from distant galaxies is red-shifted because the galaxies are moving away from us. Hubble discovered that more

Albert Einstein with astronomers at Carnegie's Santa Barbara Street library in Pasadena, 1931: (left to right) Milton L. Humason, Edwin P. Hubble, Charles E. St. John, Albert A. Michelson, Albert Einstein, William W. Campbell, Walter S. Adams. George E. Hale's portrait is in the background.

distant galaxies are moving away faster. This result is summarized in an equation known now as Hubble's law:

$$V = H \times D,$$

where V is the velocity of recession of the galaxy, D its distance, and H a number known as Hubble's constant. The current value for H is approximately 70 km per second per Megaparsec (i.e., the velocity increases by 70 km per second for every increase in distance of a million parsecs, or about three million light years).

One way to visualize the Hubble expansion is to think of the galaxies as raisins scattered throughout a piece of rising bread dough. Stand on any raisin and you will see all the other raisins being carried away from you, and the farther away from you they are, the more expanding dough will separate you and the faster their recession will be. Thus the Hubble expansion of the universe should be thought of as an expansion of space-time itself, rather than as the movement of galaxies through space.

Edwin Hubble at the 200-inch Hale telescope at Mount Palomar.

Astronomers around the world were quick to grasp the astonishing implications of Hubble's expanding universe. Such a scenario implies, for example, that the universe had a beginning at a specific time in the past (think of running a film of the expansion backward). Hubble's vision also implies that the universe will have an end, either in an infinite expansion or by an eventual recollapse. The name given to this picture of an explosive beginning followed by an expansion is the "big bang." All of modern cosmology is based on this pioneering work of Edwin Hubble. No wonder that when NASA launched the first major telescopic observatory into orbit, they named it the Hubble Space Telescope in his honor.

During the 1930s, Hubble's work brought him recognition not only among his astronomer colleagues, but by the general public as well. His image grew even larger in 1931 when Albert Einstein visited Pasadena and spent time with Hubble. Besides nurturing relationships with scientific giants like Einstein, Hubble socialized with the Hollywood set and was routinely photographed with movie stars. The publication of his popular book, *The Realm of the Nebulae*, in 1936 added to his reputation. It also revealed one of Hubble's little quirks. Despite his role as their discoverer, Hubble never referred to "galaxies." He seems to have preferred the more cumbersome term "extragalactic nebulae." It is also little known that from the 1930s on, Hubble had serious questions about the big bang scenario. He accepted his red-shift data, of course, but argued that there could be an as-yet-undiscovered process that would explain it without recourse to the idea of a universal expansion.

Hubble spent almost his entire adult life as a Carnegie staff member at Mount Wilson. In fact, it is difficult to imagine how he could have made his discoveries had he not enjoyed virtually unlimited observing time at the world's most powerful telescope. Like other Carnegie scientists, Hubble suspended his research during World War II, when he served as chief of exterior ballistics and director of the supersonic wind tunnel at Aberdeen Proving Ground in Maryland. After the war, he returned to astronomy and the work he loved. He became chairman of the Research Committee for the now-combined Mount Wilson and Palomar Observatories. The 200-inch Hale telescope at Mount Palomar, which replaced the Hooker telescope as the largest in the world, had been built jointly by Carnegie and the California Institute of Technology. It became operational in 1949, and Hubble was the first astronomer to use it for observations. Unfortunately, he worked with this instrument for only a few years. Hubble died suddenly on September 28, 1953, of a cerebral thrombosis.

Walter Baade (1893–1960)

Of the many astronomers to build on Hubble's legacy, Walter Baade stands out. Baade was born in Germany and attended Göttingen University. Because of a congenital hip problem, he did not serve in the army during World War I but combined work on aircraft design with his studies during that period, receiving his doctorate in 1919. He joined the staff at the University of Hamburg Observatory, where he remained until 1931, when he was invited to join the Carnegie Institution staff at Mount Wilson. There he worked with other astronomers in studies of distant galaxies, supernovae, and the spectra of celestial objects.

During World War II, the United States classified him as an enemy alien. Thus, while other members of the Mount Wilson staff were away working on the war effort, Baade had increased access to the telescopes. In addition, the lights of Los Angeles were dimmed at night, so the seeing on Mount Wilson was excellent. During this period Baade was able to photograph the hitherto unresolved centers of nearby galaxies. In the process, he made his most important discovery when he identified two distinct populations of galactic stars. Population I stars in the spiral arms tend to be blue, an indication of high surface temperatures. By contrast, Population II includes the brightest stars near galactic centers, which tend to be red and hence cooler. Later, astronomers realized that the Population II stars are older, while the Population I stars are relatively young.

When the Palomar Observatory opened, Baade was one of the first astronomers invited to use the new telescope. His work led to an important modification of Hubble's original estimate of the expansion rate (and therefore the age) of the universe. In 1952 he announced his results: In the collection of Cepheid variables that Hubble had used to measure the distance to galaxies, the two populations of stars had been lumped together. When stars were sorted out properly, Baade showed that the distances obtained by Hubble were actually half of their true value. This made the universe both larger and older than Hubble's data suggested.

This realization was an important step because it resolved a difficult problem that had been developing in the 1950s. The original value of the Hubble constant implied that the age of the universe was only a few billion years, but the new science of radiometric dating was giving ages close to four billion years for rocks on Earth. Had Baade not made his correction, we could easily have had a situation in which the universe was thought to be younger than some of the objects in it.

Baade retired in 1958 and spent time lecturing and visiting observatories. He died of complications following surgery on his hip. His name lives on, however. To honor his extraordinary contributions, one of the two Magellan telescopes at Las Campanas is named the Walter Baade telescope.

Allan Sandage and Wendy Freedman

Hubble's original sample of galaxies included only 22 objects, 4 of which were in a single cluster. His data displayed a lot of scatter, and many modern scientists who look at his original paper are amazed at how weak the original case for Hubble's law was. Nevertheless, numerous astronomers added data to this initial base. Today, the so-called Hubble Diagrams—graphs of velocity versus distance—can contain thousands of data points and run from the relatively nearby galaxies included in Hubble's original sample to galaxies billions of light years away. Thus Hubble's original insight, born of a flash of inspiration and scanty data, has been amply borne out. The refinement of these graphs is not an idle academic pursuit because they yield values for the Hubble constant and thus for the age of the universe. Beginning in 1976, a great debate developed over the value of the constant and the age of the universe.

The easiest way to see the relation between the Hubble constant, H, and the age of the universe is to note that a galaxy located a given distance D from Earth will be moving faster if H is large than if H is small. If H is large, it will have taken that galaxy less time to reach its location (because it is moving faster) than it would if H were small. Large values of H, then, correspond to short lifetimes for the universe, low values to longer lifetimes.

A lively debate, which the press dubbed the "Hubble wars," began when different groups of astronomers reported lifetimes of the universe that ranged from less than 10 billion years to around 20 billion years. The lower figures caught the attention of the media because, if true, the universe appeared to be younger than its oldest stars. Two astronomers at the Carnegie Observatories in Pasadena were at the center of the debate: Allan Sandage and Wendy Freedman.

Allan Sandage was born in Iowa and did his undergraduate work at the University of Illinois. He received his Ph.D. at the California Institute of Technology in 1953 and has been a staff member at the Carnegie Institution since 1952. His earliest work involved observations that proved that Walter

Allan Sandage, whose work at Mount Wilson expanded on Edwin Hubble's discoveries, 1970s.

Carnegie astronomer Wendy Freedman, 1994. Her work has refined Hubble's calculations regarding the age of the universe.

Baade's Population II stars are surprisingly old—around 12 billion years. He subsequently made important contributions to research on the age of the oldest stars, an enterprise that involves a delicate blend of theory and observation.

Born and raised in Canada, Wendy Freedman was educated at the University of Toronto, where she received her Ph.D. in 1984. During her graduate student years she began measuring the distances to nearby galaxies using Cepheid variable stars as a standard candle. She continued in this line of research after 1984 when she joined the Carnegie Observatories as a postdoctoral fellow. She quickly became a leading expert on new techniques for measuring cosmic distances, and in 1987 she was appointed a staff member at the Carnegie Observatories.

Determining the age of the universe comes down to making a precise determination of Hubble's constant. To do this, it is necessary to repeat Hubble's analysis for galaxies that are very far away, and that's where the problem starts. While it is possible to measure the red shift (and therefore the velocity) of any galaxy you can see, determining distances is much more difficult. Individual Cepheid variable stars cannot be resolved in galaxies more than about 75 million light-years away. To get a good determination of H, however, you have to be able to measure distances to galaxies hundreds of millions and even billions of light-years away. Some other, brighter standard candle had to be found.

Much of the controversy occurred as astronomers developed and refined the new standard candles. Sandage and his coworkers pioneered the use of events called Type Ia supernovae, which occur in double star systems where one of the partners is a small star known as a white dwarf. Over time, the white dwarf pulls hydrogen gas from its partner, and eventually a layer several feet thick accumulates on its surface. At this point, the tremendous pressure and temperature cause the hydrogen to flare off. Think of it as a gigantic hydrogen bomb. For several months the supernova shines more brightly than any other star, and it emits a fixed amount of radiant energy. The supernova gradually fades, but for a short time it serves as an ideal standard candle.

When a type Ia supernova suddenly appears in a distant galaxy, we know how much light it is giving off and we can use that knowledge to estimate the distance to the galaxy. In spite of technical problems in developing a distance-scale based on this idea, by the early 1990s Sandage had accumulated enough data to announce a provisional result: a value of H of about 50 km/sec/Mpc, corresponding to an age of the universe of about 20 billion years.

Wendy Freedman took a different approach to the problem. Using the unparalleled capabilities of the Hubble Space Telescope, she and her coworkers used five different methods to estimate H. Each of these methods operates according to different physical principles. For example, there is a correlation between the rate of rotation of a spiral galaxy and its intrinsic brightness so that a measured rotation rate can turn the entire galaxy into a standard candle.

Freedman's group made painstaking measurements of standard candles in 18 galaxies for which the Hubble Space Telescope could also pick out individual Cepheid variable stars. They then used these galaxies to calibrate their determination of the Hubble constant. Their method gained strength from the fact that all five standard candles gave values of H that agreed within experimental error. Freedman's preliminary estimate gave a value of H of about 80 km/sec/Mpc, and thus placed the lifetime of the universe at around 12 billion years, marginally younger than the oldest stars.

As time went on, however, both groups of astronomers improved their techniques. As more data came in, the values of the Hubble constant began to converge. At last count, both Freedman and Sandage were agreeing on an age for the universe of around 14 billion years, which is comfortably older than the oldest stars. While their values of the Hubble constant still differ by about 10 percent, the "Hubble wars" are pretty much over.

DARK MATTER

The essence of the scientific process is to make meticulous measurements. Inevitably, most data are routine; they fill in tiny gaps in our knowledge and conform to decades of expectations. But once in a great while such measurements defy all predictions and lead to a deeper understanding of the natural world. Vera Rubin's work on galactic rotations led to such a remarkable transformation.

Stargazer

When she was 10 years old, Rubin, then Vera Cooper, and her family moved to Washington, D.C., from Philadelphia, where she was born in 1928. From her bedroom window she could watch the stately wheeling of the stars in the northern sky at night. This childhood stargazing captivated her and led to a career in astronomy, a passion that has guided her life ever since. She attended Vassar College, for she knew of its history in educating women astronomers, taking her B.A. in 1948. She then went to Cornell University, where she studied physics with some of the greatest names in the field, including Richard Feynman and Hans Bethe.

*Andromeda Galaxy
(Messier 31).*

Rubin's master's research introduced her to her life's work on galactic motions. Her thesis addressed a question about the motion of galaxies: Is there, superimposed on Hubble's universal expansion, some other motion, perhaps the result of large-scale motions of groups of galaxies? Three weeks after she and her husband welcomed their first child into the world, Rubin attended a meeting of the American Astronomical Society in Haverford, Pennsylvania, where she presented her work. A newspaper headline the next day read: "Young Mother Figures Center of Creation by Star Motions."

When the young Rubin family moved to Washington, D.C., Vera enrolled at Georgetown University to begin her doctoral work. A mutual friend put her in touch with George Gamow, a prominent astrophysicist at neighboring George Washington University, and he agreed to supervise her thesis. Lacking a better meeting place, their conversations took place at Carnegie's Department of Terrestrial Magnetism at the edge of Rock Creek Park. Thinking back to those meetings, Rubin recalls, "I decided the first time I walked into the building that this is where I wanted to be." She received her Ph.D. in 1954 and remained at Georgetown as an assistant professor, studying the rotation of the Milky Way galaxy.

Image Tubes

In the meantime, the Carnegie Institution had initiated an effort that would have a profound effect on its astronomy program, as well as on programs worldwide. Observational astronomy depends on collecting light from objects in the sky. In their quest to detect ever fainter objects, astronomers in the early 1950s had run up against the technological limits of their favorite recording medium, the photographic emulsion. It took about a thousand photons to darken a single grain in the best emulsion, and this requirement set a limit on how faint an object could be and still be detected. This limitation was a serious problem because the most interesting objects in the sky are often those farthest away and therefore the faintest. Acting with the same decisiveness he had displayed during World War II, the Institution's president Vannevar Bush convened a blue-ribbon panel to investigate whether it was feasible to use new electronic technologies to create a successor to the photographic plate.

The science behind the image tube had been understood since 1905, when Albert Einstein explained the photoelectric effect. When photons of light strike some metals, electrons are ejected. The basic question faced by Bush's committee was whether those ejected electrons could be used to pinpoint the source of photons from space and, if so, could an instrument that accomplished this task be built? The committee decided that it was indeed possible but that the project should be done in conjunction with industry. With funding from the Carnegie and the National Science Foundation, work on the "image tube" began.

The final design of the Carnegie image tube involved several steps. First, the electron generated by an incoming photon was accelerated and focused by electric and magnetic fields. It was then allowed to strike a thin screen coated with phosphor (something like the screen of a computer monitor). The electron caused the phosphor to scintillate, giving off many photons, which then struck another metal to create more electrons. This amplified flood of electrons then struck a final screen, producing even more photons to be recorded by a photographic emulsion. In the resulting system, a single incoming photon could produce a blackened grain on the emulsion.

The Carnegie Image Tube Committee was inaugurated by Vannevar Bush in 1954. It evolved into a powerful collaboration among astronomers, physicists, engineers, and industry. Central to the effort was Carnegie's Department of Terrestrial Magnetism, whose director, Merle Tuve, oversaw the effort. By the time the program ended in 1970, 34 systems had been put into service at observatories worldwide. These systems transformed observational astronomy in the years before the introduction of modern charge-coupled devices, or CCD's, which produce an image that can be digitized and recorded. In its day, the image tube was state of the art—allowing existing telescopes to see farther than ever before.

When Rubin approached the Institution about joining the staff in 1965, there was a need to attract an astronomer who could help the Carnegie image tube effort reach its full potential. Rubin, along with Carnegie scientist W. Kent Ford, began observing at Lowell Observatory, in Flagstaff, Arizona, and Kitt Peak National Observatory, near Tucson. At first Rubin and Ford observed quasars, the faintest and most distant objects known, but Rubin quickly tired of the competitive atmosphere in the field. "Everybody was trying to get the biggest red shift," she remembers. Consequently, she decided to take the new instrument and apply its power to an old problem—the problem of galactic rotation curves. "It was a subject I could work on at my own schedule and astronomers would be happy to see the result," she says.

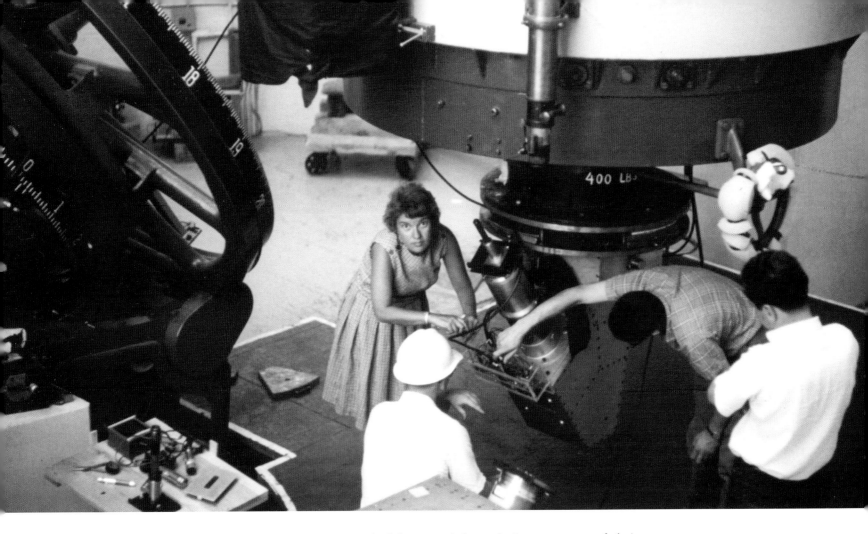

Years earlier, astronomers had discovered that galaxies rotate around their centers. The Sun, for example, makes a circuit around the Milky Way every 250 million years or so. For galaxies, we can detect the Doppler shift of light emitted by stars and deduce how fast they are going in their own circuits. (Hubble used the red shift to argue that galaxies are moving away from us.) Relative to the center of their spiral galaxy, stars on one side are moving away from us, and light from those stars and gas shows a shift toward the red spectral region. Relative to its center, the starlight from the other side of a revolving galaxy will move toward us and show a blue shift. Thus, astronomers can measure how fast a galaxy rotates and its direction.

The general expectation in the 1970s was that near a galactic center, where there were lots of stars locked together by their mutual gravitational attraction, all of them would rotate (with speed increasing with distance) as if they were a single body like Earth orbiting around the Sun. Far away from the center, on the other hand, the stars should move more slowly, just as the solar system planets move more slowly the farther they are from the Sun. This type of motion was discovered by the German mathematician Johannes Kepler (1571–1630). In a Keplerian system, the velocity of objects should decrease in proportion to the square root of the distance from the center of gravity (or axis of rotation).

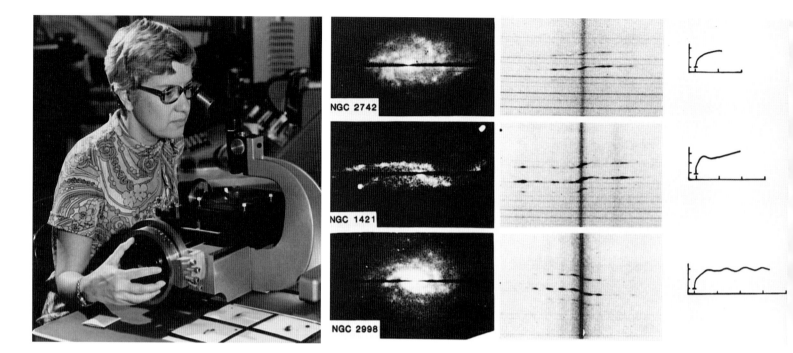

NGC 2742

NGC 1421

NGC 2998

Vera Rubin at her "measuring engine," used to examine photographic plates, 1974.

In the early 1970s, Rubin and Ford obtained rotation data for the nearby Andromeda galaxy that showed no decrease in rate even for objects at the outer visible edges. This was an unexpected result, to say the least. Rubin decided to undertake an intensive program to look at a sample of many galaxies, and to that end she began a new phase of work at Kitt Peak Observatory in 1975. It was a huge effort, but by picking galaxies farther from Earth than Andromeda, and therefore smaller in the sky, she could record three to four galaxies a night.

By 1978 the results were in. None of the galaxies showed any transition to Keplerian rotation. That is, the rate of rotation stayed constant out to the farthest stars that could be seen. Later measurements of radio waves emitted by hydrogen molecules orbiting far beyond the stars showed the same effect. As of this writing, no one anywhere has documented Keplerian rotation at the outer limits of a normal galaxy.

What do we make of this? The only possible explanation is that the luminous matter of the galaxy—stars, nebulae, dust clouds—is somehow locked into some larger invisible mass that is rotating in its own way, dragging the luminous matter with it. In a sense, the stars we see are like leaves on a stream, borne along by unseen currents.

Astronomers quickly adopted the term "dark matter" to describe the extra material whose effects Rubin had seen. To produce the observed rotation in a galaxy like the Milky Way, dark matter would have to be a flattened sphere with a mass 10 times that of the stars. Rubin's results imply that at least 90 percent of

Spectra and rotation curves for three spiral galaxies (left panels). The dark lines crossing the galaxies indicate the location of the spectroscopic slit. The step-shaped curves (center) result from the differential rotation of the spiral arms. Rotation curves (right) reveal that galactic rotation velocities remain high to large distances from the galactic center.

the mass of the universe is invisible to us, detectable only through its gravitational effect on other objects.

Actually, the concept of unseen matter goes back to a little-noted piece of work by astronomer Fritz Zwicky in 1933. Looking at clusters of galaxies, he noticed that the speeds of the galaxies were such that the cluster should fly apart. There wasn't enough gravitational "oomph" in the visible parts of the galaxies to keep the cluster together. One of two things was happening: Either what looked to us like a stable collection of galaxies was, in fact, flying apart, or there was more mass in the system than we knew—mass that we couldn't see. Zwicky believed in the second option and called this stuff "missing mass," the first historical use of this term. Most astronomers, insofar as they thought about this problem at all, preferred the first explanation. As time went by, however, more and more clusters were analyzed and found to exhibit velocities too large to hold their visible masses together. The flying-apart position became less and less tenable. Rubin's results pointed the way toward a resolution of the problem by confirming that there is more mass in the cluster than there appears to be. Astronomers today think that two masses of dark matter exist in clusters: the dark matter contributing to the mass of each individual galaxy and additional dark matter centered on the cluster's core.

So, if 90 percent or more of the universe is invisible, what is it? Some of it could be ordinary stuff, Jupiter-sized objects floating around out there and too small and dark to be seen by our telescopes. Evidence suggests that at least some dark matter is of this form. However, most scientists believe that dark matter is not ordinary stuff but some as-yet-undetected kinds of particles. Most of these theoretical objects are called weakly interacting massive particles (or WIMPs). The search for them is based on trying to detect a constant "dark matter wind" blowing through Earth that is predicted if the Milky Way is really encased in a sphere of WIMPs.

In the meantime, Rubin continues her detailed studies of the dynamics of galaxies. "It's not good for an optical astronomer to learn that most of the universe is dark," she said recently. But Rubin knows full well that learning new ideas and techniques is what science is all about. She has observing time at the Carnegie's new Walter Baade telescope, and she is enthusiastic about the opportunities this state-of-the-art observatory will offer.

"I've been looking at the sky all my life," Rubin says. "And I'm still looking."

EARTH SCIENCE

TERRESTRIAL MAGNETISM

*L*ouis Bauer was one of the Institution's earliest explorers. His plan was audacious—international in scope, multidisciplinary in subject, and open-ended in time. Bauer wanted to make a map of the variations in Earth's magnetic field. For vary the field does—and not just place to place, but year to year and even hour to hour. Bauer hoped to learn why such variations occur and ultimately to understand the origins of terrestrial magnetism and its relation to other phenomena like auroras, solar eclipses, and magnetic storms.

The Question

Variations in Earth's magnetic field posed one of the most puzzling and important scientific questions at the turn of the twentieth century. Explorers as well as scholars knew that in most places on the globe, a compass needle deviates from true north and dips downward. But seemingly unpredictable changes in the direction and strength of the field in time and space confounded scientists and navigators alike. Bauer's mission was to record these variations on a planetary scale and to discover their cause.

Louis Bauer doing fieldwork in Colombo, Ceylon, 1911.

Left: Magnetic observer Frederick Brown on camel in west-central China, 1916.

Right: Magnetic instruments carried overland in Hainan, China, 1906.

Louis Agricola Bauer was born in Cincinnati in 1865. He completed his doctoral studies on the secular (that is, the "temporal") variation of Earth's magnetic field in Berlin in 1895, whereupon he returned to the United States and pursued further studies in this mystifying area. In 1896, he published a journal, *Terrestrial Magnetism: An International Quarterly*. Three years later, in 1899, he inaugurated a modest magnetic research program at the U.S. Coast and Geodetic Survey. But when the Carnegie Institution opted to support his research in 1904, he was able to launch a coordinated global effort to tackle the question.

For the next 25 years, the Carnegie Institution's Department of Terrestrial Magnetism sponsored a virtual army of magnetic observers on land and sea. These hardy explorers had to brave subzero weather near the poles, the steam baths of the tropics, and the thin air of the Andes and Himalaya—all the while carrying heavy and delicate measuring devices.

In the first three decades of this century, departmental personnel made observations at over 6,000 stations on land, often traversing remote and difficult terrain. During an expedition in China in 1915–1916, Frederick Brown and his colleagues were accosted by bandits in Derarangai, Mongolia. They opted to travel only by night until they reached more peaceful territory.

THE AMERICAN WEEKLY

Greatest Circulation in the World

©1928, by American Weekly, Inc. Great Britain Rights Reserved.

Magazine Section of the
Washington Herald

Sunday   May 27, 1928

Magnetism, Supreme Mystery of Science

Colored Sketch of the Aurora Borealis, Reproduced from "The Forces of Nature," Published by Macmillan.

The Specially Equipped Non-Magnetic Yacht "Carnegie" Setting Out for a World Cruise in the Hope of Discovering Nature's Secrets of the Weird "Northern Lights" and the Puzzling Phenomena of the Quivering Compass Needle

NOTHING in all the fields of science is more baffling than magnetism. The quivering needle of the mariner's compass swings about as if in obedience to ghostly hands, and the awe-inspiring spectacle of the Aurora Borealis, or "Northern Lights," flashes upon man's vision and fades away without warning, while at this very moment terrific magnetic storms may be raging around him and through him, and he does not perceive them, except through the disturbance of radio or telephone or telegraph service.

The tremendous power revealed by magnetic phenomena is silent as the grave—it is neither hot nor cold; it has no taste or smell; much of it is a complete mystery to science.

The non-magnetic yacht "Carnegie" is now outfitted for a three years' tour over the seven seas in another attempt by science to solve the mystery of magnetism. It may prove the most valuable discovery ever made, or the most dangerous.

The magnetic pole is near to but not the same as the North Pole, and at present it is wandering in a northwesterly direction at about eight miles a year, but it always seems to be the centre of the Aurora Borealis disturbances. The Southern Magnetic Pole theoretically ought to be found directly opposite the Northern one, but it is far out of its proper place, which puzzles the scientists.

"True as the compass" has been a symbol of fidelity in love. Really the fickle and susceptible little instrument is true to its distant love only so long as no iron or steel is nearby. Therefore all iron and steel are eliminated from the "Carnegie" in favor of wood and non-magnetic metals.

What is it up north that excites the little steel needle?

Apparently nothing at all. The ancients thought it must be a vast deposit of magnetic ore under the Arctic ice cap, but explorers found nothing of the kind. Yet some force, too vast to be conceived in horse power, permeates the entire world. What can it be?

The next theory was that the core of the earth is a ball of iron and acts like a magnet. This seemed probable because the globe has been weighed and found heavy enough to be composed mostly of this most plentiful metal. That theory was knocked in the head by the knowledge that the interior of the earth is hot, and heat destroys the magnetism of iron.

However, it might be possible that under great pressure heated iron would retain its magnetic properties. The Geophysical Laboratory of the Carnegie Institution at Washington, under its director Dr. Arthur L. Day, tested iron under various temperatures up to 1,650 degrees Fahrenheit and pressures up to 29 tons to the square inch. They found that pressure had nothing to do with the question.

An electric current creates a magnetic field, and it is possible that the whirling of the earth's great mass generates a steady current, like the revolving dynamo in the power house. The yacht will return with volumes of dreary looking figures showing how the needle behaved at certain places and days for three years.

The Non-Magnetic Yacht "Carnegie."

But when these statistics have been digested and connected with other things, such as sun spots, earthquakes, the Aurora Borealis, radio and telegraphic disturbances, they hope to find a "key fact." This key fact may reveal the mystery, not only of magnetism, but even light and the force of gravity. Gravity, the string that holds the earth to the sun, is just as mysterious as magnetism and suspected to be only a different manifestation of the same thing. Also, many believe that the Northern Lights are caused by a sort of twanging or stretching of this string in the magnetic storms that sweep the universe.

If these scientists discover what the magnetic-gravity string is and how it works, later scientists will almost certainly show how to make use of it, just as Benjamin Franklin, fooling with a kite in a thunder storm, led the way for Edison and all electric power.

If man finds a way to relax that gravity string a little, so that the earth would swing along in its orbit at a little slower speed, he might release more power than could be gotten from all the fuel in the world.

The result of unlimited power on tap would be equivalent to making everyone a millionaire—all play and almost no work. But nobody knows what convulsions might be set up in the universe by monkeying with such a monster buzz-saw. Once loosened, the world might go plunging past the other planets and become a frozen traveller in the eternal night of interstellar space. The moon might be used as a sort of guinea pig for experiments, but even that might be disastrous. It will be like trying a parachute—it must be perfect the first try. It would be something for the entire human race to vote on before taking a chance.

Mankind may, however, prosper from something less ambitious and dangerous. For ages a considerable amount of cheap power has been obtained from the winds of the earth. If science sets up magnetic windmills and catches even the tiniest fraction of the incalculable power that rages in these magnetic tornadoes that light up the Aurora Borealis, roar into radio sets and paralyze telegraphs, it would enrich the whole world. It is not inconceivable that the "Carnegie" might be a treasure ship bringing back millions for everybody.

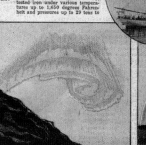

Various Manifestations of the Brilliant and Constantly Changing "Northern Lights," Which Have Been Widely Observed, Sketched, But Not Understood by Scientists.

Left: American Weekly article on the enduring "Mysteries of Magnetism," 1928.

Right: Members of the MacMillan Baffin Land Expedition with the schooner Bowdoin, 1922. Carnegie scientist R. H. Goddard appears at right.

The Institution also maintained two permanent land observatories—one in Watheroo, Australia, and the other in Huancayo, Peru. Both facilities studied a range of phenomena: terrestrial magnetism, atmospheric electricity, meteorology, Earth currents, and cosmic rays.

At Sea

It was hard enough to make magnetic measurements on land, but to do so on a pitching vessel was harder still. Yet data from the world's oceans were essential, not only for the sake of knowledge but also to correct hazardous mistakes in nautical charts.

The first research vessel sponsored by the Institution for studies on magnetism was the *Galilee,* a 328-ton brigantine chartered in 1905. Despite removal of iron fittings and steel rigging, shipboard interference with the magnetic readings remained. In response, the construction of a nonmagnetic ship was authorized.

Named the *Carnegie* after the Institution's founder, the new ship was built at Brooklyn's Tebo Yacht Basin Co. in 1909 at a cost of $115,000. The wooden

sailing vessel boasted many unusual features: copper and bronze fittings, a manganese bronze anchor with hemp cables instead of anchor chains, and a bronze engine. It was said that spinach, because of its high iron content, was stored in the engine room.

At the time of its launch in 1909, the *Carnegie* was the world's only seagoing nonmagnetic observatory. Over the next 20 years, the vessel made seven voyages of discovery. On all seven cruises, measurements of magnetism and electricity were made, but on the final cruise the mission was expanded to include studies in oceanography, meteorology, and marine biology.

The Carnegie *under sail in her trial cruise, 1909.*

Explosion of the Carnegie, *Apia, Samoa, 1929.*

Up in Smoke

The *Carnegie* exploded and burned while refueling in Samoa on November 29, 1929. In 20 years of service, the *Carnegie* had cruised 342,681 miles and gathered enough data to correct errors in nautical charts and improve understanding about the earth. The exploits of the vessel made it into print in a popular book entitled *The Last Cruise of the Carnegie*, which was written by J. Harland Paul, the ship's surgeon, and published in 1932.

Captain Ault

Captain James Percy Ault made many contributions to geomagnetic research during the first decades of the twentieth century. He participated in five of the seven cruises of the *Carnegie*, not just as commander but also as a scientist. Ault conducted land-based magnetic research too, most significantly relating to electromagnetic fluctuations during solar eclipses. He died in the explosion of the *Carnegie* in 1929.

Captain Ault in a diving helmet going over the side of the Carnegie, *1928.*

So Why Does the Magnetic Field Vary?

Despite decades of study, Bauer and his colleagues never resolved the origins of Earth's magnetic variations. It wasn't until the 1940s, long after seismologist Harold Jeffreys's 1926 discovery of the earth's liquid outer core, that convection of molten iron was implicated in these magnetic variations. Nevertheless, the Department of Terrestrial Magnetism provided a firm empirical foundation for all subsequent geomagnetic studies at a time that no other agency opted to do so. The data collected also helped improve nautical charts, which relied on magnetic measurements.

SEISMOLOGY

Since before recorded history, earthquakes have been among the most destructive of all natural phenomena. This fact was underscored shortly after the founding of Carnegie Institution when the city of San Francisco was destroyed by an earthquake and subsequent fire. That violent event inspired research that continues at the Institution to this day.

The California Problem

Seismology is the branch of science concerned with measuring and interpreting waves that travel through the earth. For example, if there is an earthquake deep underground, all sorts of waves will travel through the rock. Some of these will move through the solid earth, others along the surface. Some of the waves will be like sound waves, with the particles moving back and forth in the same direction as the wave. Others will be more like "the wave" in a sports arena, in which the particles (in this case the fans) move in a direction perpendicular to the wave itself. The theory of these waves was worked out by the great mathematical physicists of the late nineteenth century, and instruments to detect them were developed.

Santa Rosa City Hall following the "San Francisco Earthquake," 1906.

The basic detection technique exploits the fact that when one of these waves arrives at a point of the earth's surface, either via a path along the surface itself or a path through the interior, the surface will move. The presence of the wave can be detected by hanging a pendulum from a support attached to the ground. When the wave arrives, the supports of the pendulum will move with the wave, but the pendulum itself will not. This relative motion can be detected and recorded, and the recovery as the support realigns with the pendulum gives a measure of the ground motion.

To earth scientists at the turn of the last century, the great attraction of seismology was that it provided a unique way of looking at the interior of the earth and determining its internal structure. An earthquake or volcanic eruption in Japan, for example, would create waves that traveled all the way through the earth to observing stations in Europe and North America. By timing the arrivals of the different kinds of waves, scientists could reconstruct the paths that the

Damaged church in Long Beach, California, after the 1933 earthquake. The earthquake registered 6.2 on the Richter scale.

waves had followed. By applying the theory that had been developed, they could make statements about the properties of the material through which the wave had passed. All of our knowledge about Earth's interior—its iron inner core, its liquid outer core, its rocky mantle—comes from this sort of measurement. Serious seismologists in the early 1900s focused on this sort of investigation.

Then, on April 18, 1906, a massive earthquake occurred near San Francisco, and some scientists began to rethink their priorities. The earthquake and subsequent fire destroyed over 3,000 acres in the center of the city and caused at least 500 deaths. Given these losses and the clear threat of future earthquakes to do the same, it is hardly surprising that some American geologists started to move their science from its emphasis on abstract investigations of Earth's interior toward a more applied focus on earthquakes. This venture would require not only that scientists move in a new direction but that they get involved in politics. For example, after the San Francisco earthquake many local businessmen and politicians felt that talking about geological threats would be detrimental to the region's development, and thus they argued that the primary problem in San Francisco in 1906 was the fire, not the earthquake. Practical seismology, therefore, would require diplomacy and communication skills. It would also require a major new source of funding. Undaunted by such problems, Harry Oscar Wood went to work almost as soon as the smoke had cleared.

Harry Oscar Wood (1879–1958)

Harry Wood was born in Gardiner, Maine, on July 28, 1879. He attended Bowdoin College briefly but completed his education at Harvard University, where he received his B.A. in 1902 and M.A. in 1904. During his undergraduate and graduate careers, his main interest was mineralogy. Like George Ellery Hale, he never completed his Ph.D. Instead, he went west to the University of California at Berkeley in 1904 to take a position as an assistant in the geology department. Immediately after the San Francisco earthquake, he was tapped by Andrew C. Lawson, chairman of the Berkeley geology department, to help investigate the damage and make a report. The Carnegie Institution provided funds for this enormous study, which was published in 1908. By this time Wood had begun to tackle the problem of recording and predicting local earthquakes. He quickly ran into problems, both financial and technical. Because of the existing bias toward the study of the deep earth, he could not obtain the funds needed to build and maintain a network of seismic stations to map California's earthquakes. The technical problems were even trickier.

The reason that recording local earthquakes was so difficult has to do with the nature of the waves created in them. When rocks fracture and move against each other, many types of waves are generated. Some of these waves are capable of traveling long distances, and some die out quickly and don't move very far from their source. The difference has to do with the wave's frequency—how often the crest of a wave will pass a given observation point. For long-distance waves, the sort that travel through deep earth, a crest will arrive about every 5 to 10 seconds. For waves that die out quickly as they travel away from the source, however, crests arrive much more frequently, at intervals of less than 2 seconds. Because the study of long-distance waves was all that had interested seismologists, the only instruments available for detecting seismic waves were sensitive in the long range.

Trying to use them to detect the higher-frequency waves associated with nearby earthquakes is a little like trying to listen to a radio station when you are at the edge of its broadcasting range: You can do it if the conditions are right, but it's difficult and you don't always get everything.

Despite this limitation, Wood began studying local earthquakes as well as he could in California. Then, in 1912, he left the University of California (apparently there were difficulties with his teaching) and moved to Hawaii, where he became an assistant researcher at a new volcano observatory at Kilauea. There he honed his skills by measuring earthquakes associated with the movement of magma within the volcano. He also made the acquaintance of Arthur Day, director of the Carnegie Institution's Geophysical Laboratory and a visiting researcher. Later, Day would play a crucial role as Wood's mentor and supporter.

Scientists at the Seismological Laboratory, October 1929. Front row (left to right): Archie P. King, L. H. Adams, Hugo Benioff, Beno Gutenberg, Harold Jeffreys, Charles F. Richter, Arthur L. Day, Harry O. Wood, Ralph Arnold, John P. Buwalda; top row (left to right): Alden C. Waite, Perry Byerly, Harry F. Reid, John A. Anderson, Father J. P. Macelwane.

During World War I, Wood was commissioned in the Army Engineer Reserve Corps. He worked at the Bureau of Standards trying to develop ways of pinpointing the location of enemy artillery through the use of ground vibrations (a project that eventually proved unsuccessful). While in Washington, Wood lived at the Cosmos Club and began to develop a personal network that would prove valuable to him after the war. In particular, he met George Ellery Hale, who hired him after the war to be the assistant to the executive secretary of the newly formed National Research Council. Wood also re-established contact with John Merriam, who had been on the Berkeley faculty but was now serving as chairman of the National Research Council.

The time was ripe for Wood to propose his California research project again. In 1920, the National Research Council approved his proposal for a regional seismic monitoring program, and a year later the Carnegie Institution—with the hearty endorsement of Hale—decided to help fund it. In 1921, Merriam, who was now president of the Carnegie Institution, established the Carnegie Seismology Advisory Committee with Arthur Day as its chairman and Harry Wood as the research associate in charge of the day-to-day operations of the program.

Wood went to Pasadena, where, in space provided by Hale at the Mount Wilson Observatory, he began developing an instrument capable of measuring the high-frequency waves associated with local earthquakes. In 1923, after a couple of false starts, Wood, in close collaboration with Mount Wilson astronomer and optics expert John Anderson, built the first torsion seismograph. This instrument suspends a weight from two or more wires. When the support for the instrument moves during a tremor, the wires twist as they pull the weight along and then twist again, as the weight realigns itself. This rotation is optically magnified and photographically recorded by means of a mirror attached to the pendulum and the recording drum. A patent for this device was registered by the Carnegie Institution on September 1, 1925.

With his instrument in hand, Wood turned to the problem of building a set of observation stations to begin monitoring local earthquakes. Day and Merriam concurred in Wood's proposal for five stations located within 70 miles of Pasadena and arranged for the Carnegie Institution to fund them. It was at this point that Wood's connections with the Institution really began to come into play. Through Ralph Arnold, a member of Day's advisory committee, Wood was invited to speak at the Rotary Club of Riverside, California, about seismic hazards. With some prodding by Arnold, the mayor agreed to fund a building for a seismic station. T. W. Vaughn, an oceanographer who had worked on Carnegie projects, arranged for the housing of another station at Scripps Oceanographic Institution in La Jolla. Henry Pritchett, a Carnegie trustee who owned a summer home in Santa Barbara, arranged for another station to be built there on a site offered by the Santa Barbara Museum of Natural History.

Last but not least, Day approached Hale about locating the central station in Pasadena, the headquarters of the Mount Wilson Observatory. Hale pointed out that the California Institute of Technology had just received a major grant from the Carnegie Corporation to start a geology and geophysics program and proposed that the seismology work be incorporated into that department. There was some reluctance on Wood's part to turn his program over in this way, but eventually the two institutions worked out a cooperative agreement. Under the terms worked out by Merriam and Caltech president Robert Millikan in 1926, the Carnegie Institution program, under Wood's control, would be housed in the new Caltech seismology laboratory. In return, Caltech scientists would have the right to participate in the research, which would be funded by the Carnegie Institution.

Despite some minor misunderstandings—there was a comic opera flap over whose name would appear on the new research institute's stationery—the collaboration was a productive one. Wood moved into his new quarters in 1927, and by 1930 he had overseen the installation of seven Wood-Anderson seismographs in southern California. Another six recording stations were installed in other localities. Vast quantities of data were acquired and numerous new fault zones were identified by these instruments. Wood made certain that this information was communicated to the public regularly through bulletins and other publications. These efforts paid off. When the Long Beach earthquake struck on March 10, 1933, for instance, even though there was severe damage (120 dead and property damage of $400 million by today's standards), there was also a level of preparedness on the part of the inhabitants that was unprecedented. Building codes had been strengthened and emergency procedures were in place. The "California problem" was being tackled.

Although Wood was never able to establish his pet theory—namely, that weak shocks are the precursors of strong ones—he had transformed the study of seismology by promoting, then directing, this unusual collaborative venture. In 1934 he suffered a viral inflammation of the spinal cord and his leadership began to decline. But his work has been carried on by others. When Wood died in 1958, his will established a fund for the "encouragement and support of research in the geological aspects of seismology." The fund has been used to finance the Harry Oscar Wood Fellowship and the Grove Karl Gilbert Fellowship at the Carnegie Institution's Department of Terrestrial Magnetism and Geophysical Laboratory.

Even more important to the development of seismology were Wood's many collaborations at Pasadena during the first half of the twentieth century. In fact, it is probably fair to say that Wood's greatest contribution, besides implementing a regional earthquake-monitoring system, was his ability to attract extraordinary people to the program. Two gifted scientists in particular joined the program under Wood's guidance: Beno Gutenberg and Charles Richter.

Beno Gutenberg (1889–1960)

Born in Darmstadt, Germany, to a wealthy manufacturing family, Beno Gutenberg attended the University of Göttingen, where he studied mathematics and geophysics with some of the great mathematicians of Germany's greatest scientific period. He received his Ph.D. before he turned 22—an almost unheard-of feat in any university. He then turned to analyzing data on distant earthquakes and, in 1912, determined that the core of the earth begins approximately 2,000 km beneath the surface, a number very close to the currently accepted value. In 1913 he took a position at the International Seismological Association in Strasbourg, then part of Germany. After a stint as a meteorologist in World War I, he found that as a German he was no longer allowed to work in Strasbourg, which was then part of France. He returned to Darmstadt in 1919 and went to work in the family soap factory.

As things in Germany returned to normal, he started lecturing—first without pay, then after 1926 as an extraordinary (associate) professor—at the University of Frankfurt. The obvious post for him was the soon to be vacated chair in geophysics at the University of Göttingen, but Gutenberg learned through private conversations that he was unlikely to be appointed because German authorities had already concluded that there was too strong a Jewish presence in the professoriat. It was against this background that he attended a world seismology conference in Pasadena in 1929. It was an open secret at the time that Millikan was looking for a "name" geophysicist for his new department at Caltech, and Gutenberg was high on everyone's list. Harry Wood is supposed to have remarked to Arthur Day after the meeting, "We need Gutenberg more than

Beno Gutenberg studying a seismograph, 1958.

Europe does." When the rumor arose that Gutenberg was going to be offered a job at Harvard, Millikan called Wood and a few colleagues into his office. Two hours later a cable with a job offer went out to Germany.

In the words of one historian, "Gutenberg's appointment decisively shifted the center of seismological research from Germany to the United States and the center of American seismology from Berkeley to Caltech." With Wood and Richter, he analyzed seismic data—both old and new—and developed improved methods of determining epicenters of earthquakes, better ways of calculating the path of waves through the earth, and improved understanding of the earth's surface structure. He also collaborated with Richter in developing the scale that bears the latter's name.

Charles Francis Richter (1900–1985)

Charles Richter was born in Ohio, but his family moved to California when he was a teenager. He studied physics at Stanford, then completed his Ph.D. at the California Institute of Technology in 1928. Even before receiving his degree, he expressed interest in remaining at Caltech. Wood, who was impressed by Richter's unusual ability to read seismograms, contacted Arthur Day about the possibility of hiring the graduate student. At first, they discussed sending Richter to Germany to study with Beno Gutenberg, but as events played out, that proved unnecessary. Instead, Richter was added to the Carnegie staff in Pasadena, a position he held from 1927 to 1936, when he joined the faculty of Caltech. Before he left Carnegie, however, Richter developed his famous earthquake scale.

The Richter scale measures the magnitude of earthquakes. Following the lead of Japanese researchers, Richter plotted the maximum post-earthquake ground motion recorded at seismic stations as a function of the distance from the station to the epicenter. He quickly found that if he tried to do this directly, he ran into a problem. The recorded motion could be as small as 1 mm or as large as 12 cm, and there was no way to get them all on the same graph. According to Richter, he took the problem to Gutenberg, who said, "Try plotting them on the logarithmic scale." (A logarithmic scale is one where each division on the vertical axis of a graph corresponds to a factor of 10 in the quantity being measured.) Some scholars doubt this story on the grounds that it's hard to see how a theoretical physicist like Richter would need this kind of prompting. In any case, when Richter plotted things this way, an amazing correlation appeared. For each individual earthquake, the points from various stations formed a curve on the graph, and the curves for different earthquakes were all parallel to each other. This meant that by extrapolating the curves back to zero distance, he could compare the strengths (or "magnitudes") of the different earthquakes in a consistent way. In effect, this extrapolated value represents the amount of earth movement that would be felt by a seismograph located right at the epicenter of the quake.

In his original scale, Richter took a magnitude of "1" to be a shock that created an earth movement of 0.01 mm at a station 100 km from the epicenter, "2" to be a shock that produced a movement of 0.1 mm, and so on. In this way, each stage in the scale corresponds to an increase in ground motion 10 times

First overseas installation of a borehole strainmeter, Matsushiro Observatory, Japan, 1971.

larger than the previous stage. For example, a magnitude 7 quake is 100 times as strong as a magnitude 5 and 1,000 times as strong as a magnitude 4, and so on. Today, a scale similar to Richter's but based on the total energy released in an earthquake, rather than ground motion, is used.

And Then What?

The Carnegie Institution began to withdraw from the Pasadena seismology program in 1934 and by 1941 it had turned the work over to Caltech entirely. By this time, the program had expanded from one focused on local geological events in California to one that grappled with the question of earth processes, including the problem of Earth's interior that had so bedeviled Wood's earliest efforts. Today, the Seismology Laboratory continues to play a leading role in earthquake monitoring, and one often sees scientists from the laboratory quoted after major earthquakes anywhere in the world.

Seismology is alive and well at the Carnegie Institution, too, though such research in now centered at the Department of Terrestrial Magnetism. During the 1950s, Merle Tuve, director of the department and a nuclear physicist, actively sought "basic research" projects that, in Tuve's words, would lead to "greater knowledge and deeper understanding" rather than "quantity of scientific results." Seismology was one of the fields embraced by Tuve, and the department's scientists have been actively engaged in the field ever since.

One of their earliest efforts was a program in explosion seismology, an experimental program in which scientists sought to understand Earth's crust by generating seismic waves artificially through carefully located and timed explosions. Such an approach circumvented the problem of having to wait for "natural" earthquakes to occur. It had the additional advantage of putting war-surplus explosives to use. Another way of circumventing the timing problem was to do what Wood and his colleagues did in California, install arrays of seismometers that record seismic activity on a continuing basis. The development of a new generation of sensitive instruments was central to these efforts.

One such device is the borehole strainmeter, an ingenious tool that measures minute changes in the strain of rocks. In essence, the strainmeter is a metal tube filled with liquid. When buried in the ground, it detects deformation of the surrounding rock by sensing changes in the volume of the liquid in the tube. The data collected from such instruments help scientists understand the subtle crustal stress changes that lead to earthquakes. The strainmeter was developed at the Department of Terrestrial Magnetism by Selwyn Sacks, who collaborated with Dale Evertson of the Applied Research Laboratory in Austin, Texas, and staff instrument maker, Mike Seemann. The first version was tested in 1968 on the grounds of the Department of Terrestrial Magnetism. Improved models have been installed in seismically active and volcanic regions around the world. Strainmeters have played an important role in expanding our understanding of earth processes. Because they can detect "slow earthquakes," that is, earthquakes that don't register on other seismometers, they have expanded our understanding of earth dynamics. On two occasions, a strainmeter in Iceland detected a signal that preceded the eruption of the volcano Hekla. In 2000, the signal allowed a formal public prediction of the eruption.

Harry Wood's dream of predicting earthquakes has not yet been realized. But, thanks to the research program he began, scientists have a new and deeper understanding of our dynamic planet.

MEGABAR

The laws of nature restrict human access to space and time. There are feats scientists can never accomplish, places they can never visit, and questions they can never fully answer. Earth scientists, in particular, study the planet under these daunting limitations. On the basis of current knowledge about the strengths of materials and the hazards of extreme temperature and pressure, journeying even a dozen miles into the crust is beyond any plausible technology. How, then, can scientists know what forms Earth's deep interior?

Seismologists learn much about the structure of Earth's layered interior through indirect observations of sound waves, which help to document the density and mechanical properties of earth materials. But scientists can't understand the nature of planetary interiors without first subjecting rocks to extreme conditions of pressure and temperature and measuring how they behave. Most portions of the earth's interior experience pressures of up to 3.5 million atmospheres—pressures that force atoms into new, more densely packed arrangements than those found on the surface of the planet. Properties of minerals depend on their atomic structures, so an understanding of Earth's core and lowest mantle regions requires an intimate knowledge of the effects of pressures of a million atmospheres and more.

In 1903, a blue-ribbon panel of earth scientists presented the Carnegie Institution's trustees with a comprehensive prospectus for "geophysical" research on the physical and chemical properties of rocks and minerals. Prominent among their recommendations was an extensive series of experiments employing "high pressure in closed vessels." With this proposal in hand, the Institution authorized the construction of a Geophysical Laboratory in 1905, and from the outset, high-pressure research became a central focus of the new facility.

The Heroic Age of High-Pressure Research

Pressure is defined as a force applied to an area. For most of the twentieth century, high pressures were generated in massive steel squeezers that applied tons of deadweight force to modest areas. Many of these epic devices pressed hardened metal pistons into rigid, sample-filled cylinders. Other apparati employed compressed gas or liquid as a pressure medium. These were dangerous experiments, and every laboratory that pushed the limits of pressure experienced its share of explosive failures.

At the turn of the twentieth century the most challenging part of squeezing matter was confining the sample so that it didn't squirt out. Fortunately for the planners of the Geophysical Laboratory, that problem was about to be solved by Harvard physicist Percy Bridgman (1882–1961). Prior to Bridgman's work, experimentalists rarely achieved sustained pressures greater than 2,000 atmospheres. In about 1905, by some combination of luck, intuition, and keen observation, Bridgman devised a self-sealing pressure chamber that permitted much higher pressures. With this invention, Bridgman proclaimed, "The whole high-pressure field opened up almost at once before me, like a vision of a promised land." By 1910, Bridgman was routinely subjecting samples to 20,000 atmospheres. He pointed the way for Carnegie's earth scientists.

During the laboratory's first two decades, dozens of high-pressure studies on mineral and rock properties were published by laboratory scientists. Arthur L. Day, an early recipient of an individual Carnegie grant and the Geophysical Laboratory's first director, studied effects of pressure on mineral melting as early as 1904 and established a high-pressure laboratory in the new building.

The most influential high-pressure work of the laboratory's early years was conducted by petrologist Norman Levi Bowen, who formulated a comprehensive theory of the origin of igneous rocks. Bowen was a graduate student at the Massachusetts Institute of Technology in 1909 when Arthur Day invited him to conduct his thesis research at the Geophysical Laboratory. Upon graduating in 1912, Bowen joined the laboratory's staff, a position he held for most of the next half-century. Bowen and his colleagues focused on effects of temperature and pressure on the stability and melting of rocks. This work culminated in 1928 with his landmark book, *The Evolution of the Igneous Rocks*. Bowen argued persuasively that when a rock partially melts, the compositions of the melt and the left-over minerals may differ significantly. Repeated cycles of partial rock melting and extraction of this melt, therefore, can produce the wide variety of observed igneous rock compositions.

High-pressure research was all but halted during World War II, but the Geophysical Laboratory's experience in constructing cylindrical steel pressure vessels was put to good use in the development of improved machine-gun barrels. The high-pressure program continued in earnest after the war with new research on the origin of igneous rocks. In the mid-1950s, petrologist Francis R. (Joe) Boyd and his longtime colleague, Joe England, used an improved piston–cylinder apparatus to tackle pressures approaching 100,000 atmospheres. They published numerous studies of the stability and behavior of rocks from Earth's upper mantle.

Bowen's classic studies on the evolution of the igneous rocks were augmented by the high-pressure work of Hatten S. Yoder, Jr., who designed a massive 12-inch-diameter steel cylinder that could sustain gas pressures of 12,000 atmospheres at temperatures approaching 2,000°C. Yoder, for the first time anywhere, was able to duplicate conditions throughout the entire crust of the earth, and he used this machine to tackle a classic problem, the origin of basalt. With his enhanced pressure capability Yoder was able to demonstrate that many different varieties of basalt could be explained as the result of partial melting of similar rocks at different depths.

Under Pressure

Pressure at Earth's core exceeds 3 million atmospheres, but before the 1970s most geoscientists were content to focus on the much lower pressures that are found in the outer few hundred miles of the planet. With so much to learn about rocks at thousands of atmospheres, there seemed little urgency to reach for a million or more.

Priorities changed in 1968 when a young high-pressure scientist, Ho-Kwang ("Dave") Mao, accepted a position to work with staff scientist Peter M. Bell. It would be hard to imagine two more different scientists. Mao, the son of a Nationalist Chinese general, was born in Shanghai in 1941 but grew up in Taiwan after his family fled the mainland with the defeated Nationalists. His remarkable career in high-pressure research came about by pure chance. All students enrolling in Taiwan's university system listed their choices for school and subject major and then took a difficult six-part entrance examination. The highest scorers received their first selections. Mao placed in the top 5 percent, but by the time his score came up, his top choice, the physics department at Taiwanese National University, was filled. In fact, he had to settle for the university's geology department, his eighth pick. One point higher on the 600-point examination and Mao would have studied food production as an agricultural chemist; one point lower and he would have become a veterinarian.

After graduation in 1963 and a year as a second lieutenant in the Taiwanese Air Force, Mao came to the United States for graduate studies. He enrolled at the University of Rochester because it had no application fee. It was at Rochester that Mao mastered an entirely new kind of pressure device, the diamond-anvil cell.

The diamond-anvil cell employs a simple, elegant design devised by Percy Bridgman in the 1930s. Bridgman achieved record pressures greater than 100,000 atmospheres by inserting two tapered steel or carbide anvils in a large press and squeezing them together. In 1959, scientists working independently at the National Bureau of Standards in Washington, D.C., and at the University of Chicago achieved a similar effect using much smaller diamond anvils. This remarkable new device was championed by National Bureau of Standards scientist Alvin Van Valkenburg, who first looked through diamond into the world of high pressure. By placing a small metal gasket between the diamonds and placing liquid into the gasket hole, he confined the liquid at pressure. Van Valkenburg also showed how to surround a tiny crystal with a liquid pressure medium so that the crystal would not be crushed. With this improvement, Van Valkenburg became the first person to watch as minerals changed colors, chemicals reacted, and water froze into new crystal forms at high pressure.

Among Van Valkenburg's diamond-cell disciples was William Bassett, an assistant professor of geology at the University of Rochester. Bassett and fellow

Peter Bell and Ho-Kwang Mao (right) calibrating pressures prior to an experiment at the Geophysical Laboratory, 1978.

faculty member Taro Takahashi became David Mao's Ph.D. thesis advisers in the first study of the properties of iron, the principal element of the earth's core, at high temperatures and at pressures greater than 100,000 atmospheres. This extraordinary effort caught the eye of Carnegie scientists, and upon graduation, Mao was offered a postdoctoral fellowship at the Geophysical Laboratory. There he began a 20-year collaboration with Peter Bell.

Bell was born in 1934 in New York City but grew up in exotic ports in Trinidad, the Bahamas, and Venezuela, where his father served as an executive in a shipping company. His education included exclusive New England prep schools and culminated with a doctorate in geophysics from Harvard, where he too was influenced by Bridgman.

It's not always easy to analyze what makes a successful collaboration, but the groundbreaking contributions of Mao and Bell resulted, at least in part, from their complementary scientific skills. Mao knew the diamond cell as well as anyone. Bell was an expert in big presses and had a broad background in earth and planetary physics. Mao wanted nothing more than a quiet laboratory and the time to perform experiments. An equally avid experimentalist, Bell also excelled on the public front, giving talks, securing grants, and welcoming visitors to the lab. Both men were exceptionally creative thinkers; together they worked high-pressure magic.

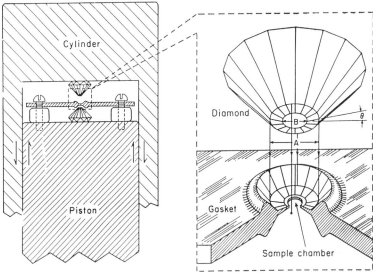

If Mao and Bell had the desire and ability, the Carnegie Institution gave them the opportunity. When Mao joined the Institution, the diamond-anvil cell created by Van Valkenburg and his colleagues at the National Bureau of Standards (NBS) had remained almost unchanged for more than a decade. At the routine 300,000-atmosphere operating range of the cell, samples were already almost too small to measure. Higher pressures would mean even smaller samples, which were beyond the capabilities of available instruments.

The development in the early 1970s of high-intensity X-ray sources and powerful lasers provided scientists with new tools to probe microscopic samples. Mao adapted these techniques in his quest for higher pressures. In the NBS cell design, for instance, pressure was limited by the relatively soft steel that supported the diamonds. Above a few hundred thousand atmospheres, the diamonds sank into the deformed steel support and no further pressure increase was possible. Mao reasoned that a harder "backing" material such as tungsten carbide would increase the pressure range, but he also knew that direct diamond-to-diamond contact at those pressures would almost certainly cause diamond breakage. Van Valkenburg's metal-gasketing technique provided an elegant solution to the dilemma. As diamond anvils deform a gasket, the metal automatically behaves like a binding ring, providing additional support to the stressed diamond tip.

Perfect alignment of the diamonds was another critical factor in reaching a pressure record. The two diamond-anvil faces had to be exactly parallel or the gasket would deform unevenly and catastrophically, squirting the sample out sideways. The two anvils also had to approach each other along the same central axis to maintain stability as pressure increased. To accomplish that difficult mechanical feat, Mao and Bell devised a special piston–cylinder device to hold the diamonds.

Left: Opposed diamond anvils form the heart of the diamond-anvil high-pressure cell.

Right: Cross section of diamond-anvil cell.

To perfect their diamond cell, Mao and Bell engaged in a long process of trial and error, breaking expensive diamond after diamond. Dave Mao remembers the all-too-familiar sickening snap, the muted but distinct sound of yet another diamond being destroyed by pressure. Eventually, however, they hit upon a felicitous combination of components that permitted much larger samples and much higher pressures. With this improved design, Mao and Bell set their sights on a new high-pressure record: a million atmospheres, or a megabar. Several scientists had "proven" on theoretical grounds that a megabar was physically impossible with any known material. Mao and Bell thought differently; they were determined to sustain a megabar and measure samples at that extreme.

For their first experiments they decided to crush ruby, whose characteristic brilliant red fluorescence when excited by a blue laser beam would serve as their megabar marker. Mao and Bell knew that a wavelength shift of 37 units in ruby fluorescence corresponded to an increase of 100,000 atmospheres in pressure. Therefore, they needed to observe a total ruby shift of about 370 units to stake the megabar claim.

The critical experiment took place just before Christmas of 1975, during the Geophysical Laboratory's annual holiday vacation. The lab building was quiet as they worked through the night. The newly modified diamond cell held its compressed cargo of crushed ruby, ready to fluoresce and reveal the changing pressure. The first turns of the screw were routine: a 30-unit shift meant 83,000 atmospheres; then 45 more units to 206,000 atmospheres. It took only about 20 minutes to make and measure each pressure step. Three more increments took them to 619,000 atmospheres. Now each turn of the screw brought the nervous scientists a new world-record pressure. With another twist the pressure shot up to nearly 800,000 atmospheres, with a 290-unit shift in the ruby fluorescence line, but the next turn brought only a slight rise, to 820,000 atmospheres. Mao feared that diamond failure was near, that the cell could take no more and had begun to bend. It was literally a megabar or bust, and with another two twists they did it. The ruby line had shifted a record 370 units. They'd achieved—and documented—a pressure of 1,018,000 atmospheres.

On December 29, 1975, the two researchers submitted a terse 800-word report to *Science*. Staking their claim to the 1-megabar mark, they described "the highest pressure ever reported for a static experiment in which an internal standard was employed." Their historic article appeared just two months later.

The New Periodic Table

The magical power of pressure lies in its ability to transform matter. Under high pressure ordinary gases become liquids, everyday liquids become solids, and mundane solids become shiny metals. At 1 megabar, every element of the periodic table transforms to a new crystalline form. Every element displays new chemical and physical properties. If the periodic table is "the chemists' playground," then high-pressure scientists are showing us extraordinary new ways to play the chemists' game.

Achieving the pressure of a megabar was a satisfying record, but the real scientific work was just beginning. One of the biggest scientific prizes lay in transforming the simplest gas, hydrogen. Physicists predict that if you squeeze hydrogen gas enough it will become a solid metal, one of the strangest substances imagined by scientists. It could be a superconductor, a material that transmits electricity with no loss of electrical energy, at room temperature. Theorists also calculate that metallic hydrogen could be the most concentrated possible form of chemical energy, a potent new source of rocket fuel and explosives. Yet in spite of these possible applications, most physicists view high-pressure hydrogen research as a crucial test of theory. Every new measurement on hydrogen constrains models of matter. Space scientists see another payoff: The giant planets of our solar system—Jupiter, Saturn, Uranus, and Neptune—consist almost entirely of pressurized hydrogen. The formation of metallic hydrogen thus became a focus of high-pressure research during the last quarter of the twentieth century.

In 1979, Mao and Bell's diamond cell held the world's static high-pressure record of 1.7 megabars. They were poised to take hydrogen to a megabar and beyond. Unfortunately, squeezing hydrogen gas is vastly more complicated than squeezing powdered minerals. The first approach was to cool the hydrogen to −253°C, creating a liquid a thousand times denser than normal gas. Adapting well-established methods for handling liquefied gases, Mao and Bell loaded their diamond cell into a 6-foot-tall thermos bottle. They poured ultracold liquid hydrogen into the cell, sealed it, and with a turn of the screw hydrogen was compressed to a record 50,000 atmospheres. The element changed from a colorless liquid to a colorless icelike solid, and Mao and Bell became the first humans to look at solid pressurized hydrogen at room temperature. But to produce hydrogen in a metallic state would require much higher pressures.

In an effort to simplify the operation, Mao learned how to load a diamond cell at room temperature with 2,000 atmospheres of pressurized gas. Lab scientists loaded more than a hundred cells with gases like argon, neon, and methane without a hitch. The first hydrogen loading, however, was a different matter. Protected by an armored barrier, Mao started to pump hydrogen gas into the container. Suddenly the building shuddered as the thick steel cylinder exploded, propelling a fist-sized metal chunk across the room and sending out a ricocheting shower of shrapnel. The noise deafened Mao, who had been standing in the projectiles' paths moments before. "I was so shaken that I almost had a car accident going home," Mao remembers.

Despite many setbacks, Mao doggedly pursued the hydrogen problem. Armed with improved loading techniques, Mao and Russell Hemley, who joined the effort in 1984, eventually squeezed hydrogen to above the 2-megabar mark. At a pressure estimated to be greater than 2.5 megabars, the hydrogen began to darken, one of the first signs of possible metallization of this ubiquitous gas. The story of the unusual darkening of hydrogen under pressure made headlines around the world.

Mao, Hemley, and their many coworkers continue to pressurize a wide range of materials, including the minerals and metals of planetary interiors, high-tech superconductors, and crystallized gases. They have discovered new high-pressure forms of ice, they have documented dense high-pressure atomic arrangements in many elements, and they have investigated effects of pressure on the electronic and magnetic properties of many novel materials. Yet, while the materials and techniques have changed and the range of pressures has increased 1,000-fold, the lure of high pressure remains much the same as it did for Adams, Bowen, and other Carnegie scientists, all of whom wanted to discover how the world works beyond the relatively benign realm of the earth's surface.

LIFE SCIENCES

HEREDITY

*I*n 1902, the mechanisms of heredity represented one of the greatest scientific challenges. The groundbreaking hybridization research of Gregor Mendel, which postulated the existence of "atoms of inheritance" (what we call genes), had been rediscovered in 1900. But scientists did not know the chemical nature of genes, nor could they deduce the mechanisms by which genes convey genetic information. These grand questions have provided a central focus for almost a century of biological research at the Carnegie Institution.

The first board of trustees had to decide not only what research to support but also how to organize that support. Andrew Carnegie preferred that the Institution fund scholars wherever they worked rather than establish its own laboratories. In the early years, as mentioned previously, the Institution did both—occasionally in the same field. At the Institution's Station for Experimental Evolution on Long Island, for instance, a succession of scientists tried to understand how traits are passed from parent to offspring in a variety of organisms. But lasting and fundamental understanding of inheritance came from another Carnegie scientist—one who worked in university laboratories. His name is Thomas Hunt Morgan and he received institutional support for more than a quarter of a century.

"The Boss." Thomas Hunt Morgan in the fly-room at Columbia University, 1917.

Flies, Genes, and Thomas Hunt Morgan (1866–1945)

Thomas Hunt Morgan was born and raised in Kentucky in a family intimately tied to some of the most exciting events of the nineteenth century. His diplomat father had been American consul in Sicily and had helped Garibaldi in his quest to unify Italy. His uncle was a general in the Confederate Army and the leader of Morgan's Raiders, while his great grandfather was Francis Scott Key, composer of "The Star-Spangled Banner." Morgan spent much of his boyhood roaming the Kentucky hills collecting fossils and later worked for the U.S. Geological Survey. After graduating with a degree in zoology from what is now the University of Kentucky, he went to the Johns Hopkins University for graduate work on marine organisms.

At Johns Hopkins, Morgan studied sea spiders, particularly their embryonic development. During this period he considered himself a morphologist—a scientist who tried to understand the evolutionary relationship between organisms by studying the current properties of those organisms. But he grew disillusioned with the traditions of classical biology and their concentration on description and classification. He began to realize that biological organisms function in accordance with the ordinary laws of physics and chemistry and that the way to understand their functioning was through the same kind of experimental program that had led to progress in the physical sciences.

In 1891 Morgan joined the faculty of Bryn Mawr College, where he examined the factors that influence the development of embryos. At the time, biologists were debating the process by which a single fertilized cell grows into an adult: Was the process governed by internal factors (what we would call today "genetic" factors) or by external or environmental factors? Embryos of marine organisms, for example, could be strongly affected by the salinity of their environments. Could this sort of environmental influence explain all of an embryo's development, or was there some inherent factor at work? This was the state of the field when Morgan entered the picture.

From his studies, which continued after he moved to Columbia University in 1904, Morgan began to realize that the embryo seemed to display a predetermined tendency to reach its goal and that environmental factors could influence development only within certain well-defined limits. Nevertheless, at the time he was a vocal critic of Mendelian genetics and scoffed at the American plant breeders

PLATE 5

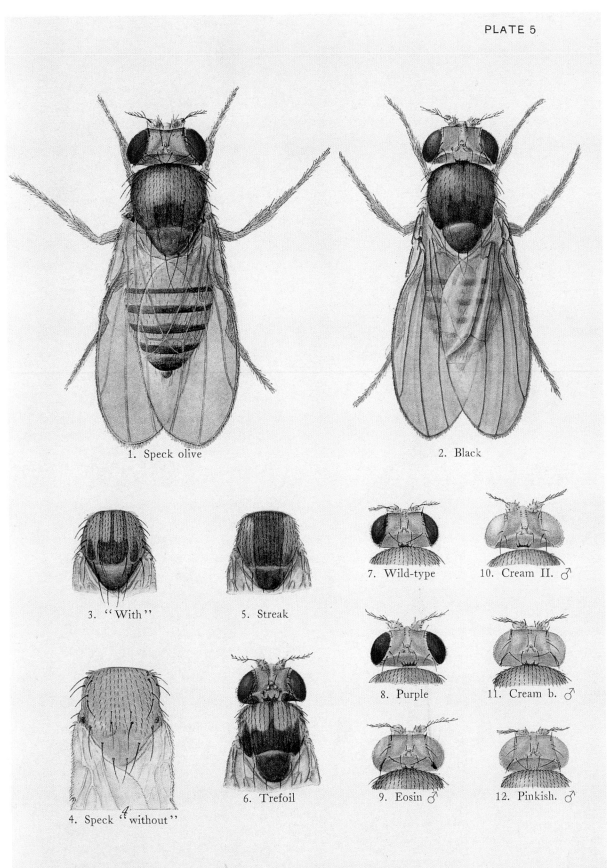

Second chromosome mutants of Drosophila, *watercolor illustration by Edith Wallace, published in CIW publication #278,* Contributions to the Genetics of Drosophila melanogaster, *1919.*

1. Speck olive

2. Black

3. "With"

5. Streak

7. Wild-type

10. Cream II. ♂

4. Speck "without"

6. Trefoil

8. Purple

11. Cream b. ♂

9. Eosin ♂

12. Pinkish. ♂

E. M. WALLACE Pinx

SECOND CHROMOSOME MUTANTS OF DROSOPHILA

who had quickly confirmed and extended Mendel's concepts soon after their rediscovery by European scientists in 1900. Even though he was critical of the environmentalist school of thought, he was clearly opposed to the notion that there was something in the fertilized egg that determined how it would develop. His own explorations of the basis for sex determination convinced him to accept only that the fertilized egg might have an inherited predisposition toward either maleness or femaleness. Morgan chafed at the philosophical, almost speculative nature of the discussion about genes. For a man who believed deeply that biology was supposed to be an experimental science, the notion that something as important as heredity and development could be controlled by unseen and perhaps unseeable objects was simply anathema.

Sometime before 1909, Morgan started doing experiments with the fruit fly, *Drosophila melanogaster*. In doing so, he was joining a small but growing group of investigators who were exploiting the properties of this organism. *Drosophila* was a particularly useful animal for Morgan's studies because it can be grown

Left: Alfred "Hot Dog" Sturtevant was inspired by Morgan's work and joined him in the "fly room."

Right: Calvin B. Bridges, one of Morgan's best students, worked briefly at Carnegie's Department of Genetics and was a Carnegie staff member from 1919 until his death in 1938.

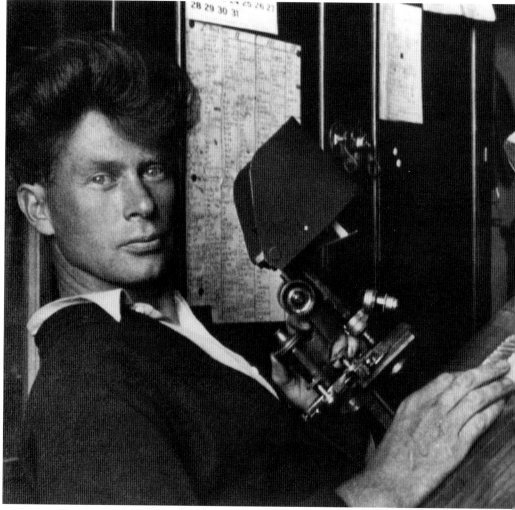

quickly in small bottles, eating only the fungi found on rotting fruit like bananas. It also produces many offspring with a breeding cycle of only about four weeks so that many generations can be studied in a relatively short time.

When Morgan began working with *Drosophila,* he had an eye on discovering mutations that would provide a system for experimental evolution. For two years his work led nowhere. Then, early in 1910, various mutant flies began appearing in his bottles. That spring he discovered a variant male fly that had white eyes instead of the red eyes typical of *Drosophila.* Breeding this fly and its offspring through several generations, Morgan discovered that all the white-eyed offspring were males and they had inherited the trait from their red-eyed mothers. Moreover, the ratios of red- to white-eyed male flies was exactly what would have been predicted by Mendelian genetics if the white trait was associated with a mutant version of a gene carried on the X chromosome. Males have only one X chromosome, which they inherit from their mothers, so any X-associated mutation shows up. In females, which have two X chromosomes, the mutation is masked by the "good" gene on the second X. This discovery, and others that quickly followed, convinced Morgan that Mendel had had the right idea. Morgan, a true scientist, was ready to abandon decades of firmly held and published beliefs when the data demanded it. By the time he applied to the Carnegie Institution for research support in 1914, he and his Columbia University students were well on their way to establishing the chromosomal theory of heredity. The "fly room" at Columbia was the bright center of genetic research.

Two brilliant undergraduate students, Alfred Sturtevant and Calvin Bridges, were inspired by Morgan's classroom lectures at Columbia and joined him in this work. Eventually, they obtained their Ph.D. degrees with him. This threesome would stay together for decades. Much of what they learned from their experiments remains fundamental to genetics today. They determined that groups of genes are joined on what they called "linkage groups," which they correctly surmised were the individual chromosomes observed in the microscope. They also found that genes and mutant versions of genes that are linked on one chromosome usually remain together as the chromosomes are passed from one generation to another. Furthermore, the linkages are occasionally unstable and the frequency at which the linkage between two genes is broken differs depending on the particular pair of genes being studied. They argued that the disruption of linkages resulted from an exchange of segments between homologous chromosomes inherited from each parent, a process called genetic recombination.

One day, while Sturtevant was talking with Morgan, the undergraduate realized that the frequency of recombination provided a direct measure of the relative distance between genes in a linkage group. A simple calculation would allow him to determine how far apart linked genes were located on a chromosome as long as he assumed that the exchange points were randomly determined. The idea was this: two genes that were close to one another were likely to be included together on a chromosome segment that was exchanged and the offspring of the flies would carry the traits associated with both of the genes. Contrarily, if the genes were located far apart, it was more likely that they would separate when the exchange occurred, so that offspring with only one or the other of two original traits would be seen. The next day, Sturtevant arrived in the fly room with the first genetic map of a chromosome.

With this work, Morgan and his group established that genes are laid out on chromosomes in a linear and knowable order. In a sense, the current Human Genome Project, with its detailed mapping and sequencing of human DNA, can be said to have started in that conversation at Columbia in 1911. Later, in the early 1930s, Barbara McClintock and Curt Stern proved definitively that genetic recombination involves the actual exchange of chromosomal segments.

This story illustrates another aspect of Morgan's work. Everyone who worked in his laboratory commented on the egalitarian nature of the enterprise. Ideas flew around freely, everyone knew what everyone else was doing, and each scientist helped his or her colleagues. Morgan was no distant "Herr Professor," but rather a man who listened to anyone who had a good idea, even an undergraduate. Over the years, this approach has become the style in many of America's best research laboratories, but Morgan was one of its first practitioners.

The filthy, crowded fly room at Columbia became the mecca for any scientist wanting to learn genetics and to be at the center of discovery. The Carnegie Institution, recognizing the extraordinary significance of the work, promptly responded to Morgan's request for funds in 1914. Its support for his research group continued when Morgan went to the California Institute of Technology in 1928 to organize the new division of biology at the invitation of George Ellery Hale. Funding ceased only at Morgan's retirement in 1942.

In Pasadena, Bridges remained a Carnegie employee while Sturtevant became a Caltech professor. Morgan created an organization based on his strongly held conviction that biology had to become an experimental science. In his personal research, he returned to the question of embryonic development and left *Drosophila* genetics to Sturtevant, Bridges, and others. But the personal and intellectual relationship with his lifelong colleagues remained close. When Morgan received the Nobel Prize in 1933, for example, he shared the prize money with Sturtevant and Bridges so that they could pay their children's college tuition bills.

Maize, Jumping Genes, and Barbara McClintock (1902–1992)

Shortly after Vannevar Bush became the president of the Carnegie Institution in 1939, he appointed Milislav Demerec director of the Department of Genetics. Demerec, a longtime Carnegie staff member, was a maize and *Drosophila* geneticist in the Morgan tradition.

One of Demerec's first actions was to hire Barbara McClintock as a staff member. She had been raised in Brooklyn, New York, and attended Cornell University, receiving her Ph.D. in 1927 in plant genetics. After several postdoctoral fellowships and junior faculty appointments, during which she made

Barbara McClintock as a student in botany at Cornell University, 1927.

extraordinary contributions to genetics, she was still without a permanent job. She had initiated a rigorous approach to cytogenetics, the marriage of traditional genetic analysis to the structure and function of chromosomes as seen in a microscope. She had proved that genetic recombination reflects the physical exchange of chromosome segments during the formation of germ cells. She had discovered and named the important nucleolar organizer regions of chromosomes. Before too long she would be elected president of the Genetics Society of America and a member of the National Academy of Sciences, only the third woman to be so honored. But in those days, academic institutions did not usually appoint women professors and McClintock's independent, feisty spirit did not fit in well at traditional institutions. Carnegie was the right place for McClintock and she remained a Carnegie scientist at Cold Spring Harbor until her death in 1992 at age 90, long after the Carnegie Institution withdrew support of that site.

While McClintock was still an assistant professor at the University of Missouri she began a long-term study of the behavior of maize chromosomes. This work was continued after she arrived at the Carnegie Laboratory at Cold Spring Harbor. McClintock was not the first maize geneticist at Cold Spring Harbor. Early in the century, George Shull had helped launch a revolution in corn production by his work on hybrids.

McClintock established that under certain conditions, the chromosomes in the cells of a maize plant will break, and, since cells do not tolerate broken chromosomes, the loose ends quickly fuse with other broken chromosomes, only to break again. Furthermore, this cycle of breakage and fusion persists through many cell divisions.

McClintock found that some patterns of chromosomal breakage were associated with changes in the pattern of pigmentation of the kernels. Eventually these studies led her to realize that some bits of chromosomes could be detached from their original positions and reattached elsewhere. It was these "controlling" elements as she called them, what are now known as "transposons" or "jumping genes," that were McClintock's greatest discovery and the justification for the Nobel Prize she received in 1983.

Here's an example of how a transposon might work: Suppose that a plant has an active gene that is required for the formation of a molecule that gives a deep purple color to all the plant's kernels. According to expectation in the 1930s, the plants that grow from such kernels should also have purple kernels unless a mutation that inactivates that gene occurs. Such mutations do occur, with the result

Barbara McClintock, ca. 1983.

Corn produced for Barbara McClintock's landmark experiments on the genetics of maize

One of Barbara McClintock's images of corn chromosomes. Shows chromosomes during prophase of meiosis. The large round structure is the nucleolus, which is attached at a specific locus to one of the chromosomes.

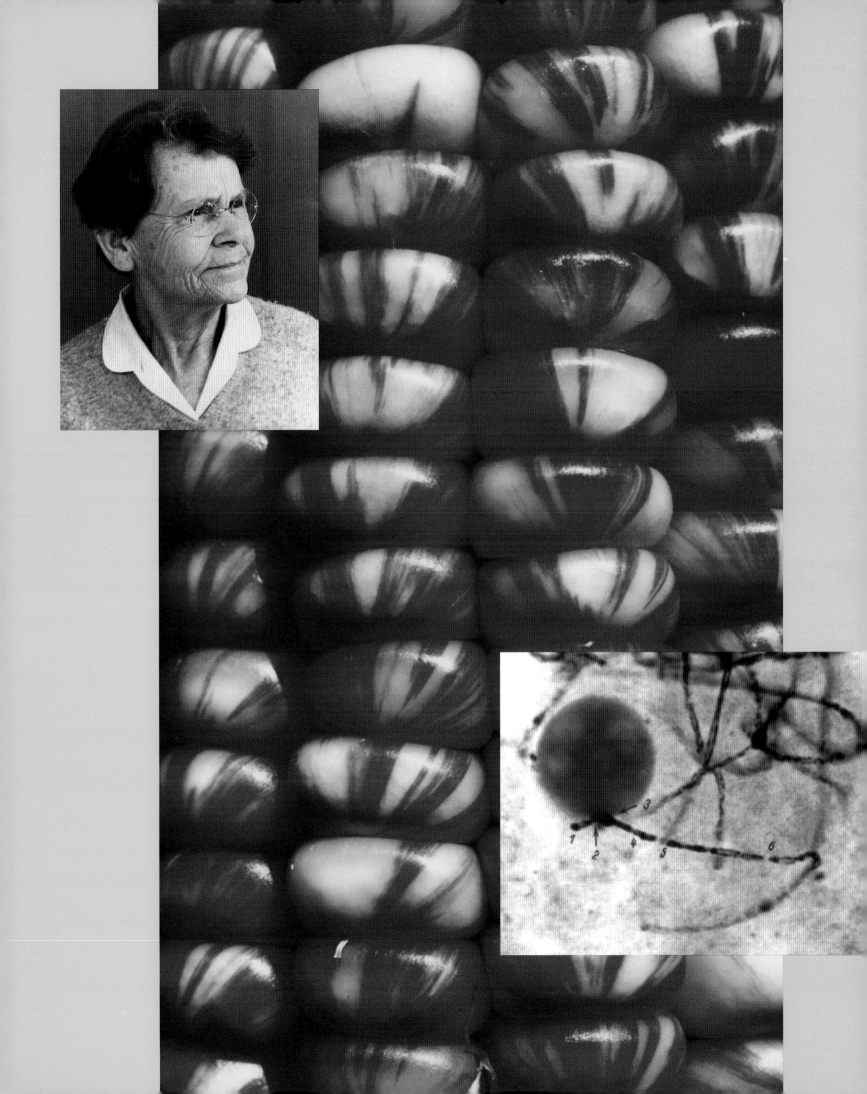

that the kernels are, for example, white rather than purple. Again, the plants that grow from white kernels are themselves expected to have white kernels, because mutations are very rarely reversible. However, if the mutation is caused by the disruption of the gene by insertion of a DNA segment from elsewhere in the genome—a transposon—the mutation can be reversed more frequently. This occurs because the transposon can also move out of a gene, restoring its original function.

What McClintock observed was that instead of the kernels being all white, they had patches of purple color. She recognized that these arose from individual kernel cells that had lost the transposon. When such cells with a now-functional gene gave rise to daughter cells, all the daughter cells were purple and formed a purple patch. Those parts of the kernel built from cells still containing the transposon remain white.

It wasn't until 1951 that McClintock had worked out the details of her theory and accumulated all of the results she needed to back it up. She reported her work at the major annual symposium at Cold Spring Harbor. Her presentation has entered scientific folklore as a classic case of a visionary, ahead of her time, being unappreciated by her mainstream colleagues. In the words of one commentator: "Few understood what she was saying and fewer still were prepared to accept it. Some, I am told, simply failed to see how one person could possibly have done all the work necessary to establish the conclusions."

With the exception of maize geneticists, few people paid much attention, but McClintock plowed ahead gathering more and more data consistent with the existence of transposons. Frustrated by the lack of interest in her papers in standard scientific journals, she published her results in the annual reports of the Carnegie Institution—surely some of the most groundbreaking contributions to that series of publications.

Eventually, though, science caught up with her. Transposons were found not only in maize, but in bacteria, fruit flies, and even humans—in short, in all organisms where anyone looked for them. Modern geneticists, who think of genes as segments of DNA rather than abstract segments of chromosomes, have confirmed through biochemical analysis that segments of DNA do indeed "jump," as McClintock claimed. In the later years of her life, honors came thick and fast, although she regarded such things as distractions from her work. Having no telephone, she heard about the Nobel Prize on her home radio and is supposed to have said, "Oh, dear," before going out for her customary morning walk in the Cold Spring Harbor woods. Upon receiving the Prize, she remarked,

"It might seem unfair . . . to reward a person for having so much pleasure, over the years, asking the maize plant to solve specific problems and then watching its response."

Barbara McClintock's Legacy

McClintock's work was carried on by Nina Fedoroff, who helped to confirm McClintock's conclusions at the level of DNA structure. Appointed a staff member at Carnegie's Department of Embryology in 1978, Fedoroff shifted her focus from research on frogs, which she had pursued in Donald Brown's laboratory at the department, to studies of maize transposons using the tools of molecular biology. Her first seeds came from McClintock's prized collection. She used molecular cloning to obtain pure transposon DNA so that its detailed structure could be determined and the mechanism by which it inserts itself into and is ejected from unrelated DNA segments could be studied.

Transposons also provided the research focus for Maxine Singer, who became president of the Carnegie Institution in 1988. Singer discovered that the dispersed repeated segments in the human genome were copies and remnants of a jumping

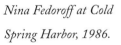

Nina Fedoroff at Cold Spring Harbor, 1986.

gene. In humans, as in maize and flies, transposons cause mutations when they land in a gene. These mutations can result in disease. But in laboratories, transposons can be used to generate mutations in model organisms—an important tool for understanding what individual genes do.

Allan Spradling and Gerald Rubin of the Department of Embryology, working with *Drosophila,* were the first researchers to develop this tool. Almost overnight, the task of finding and isolating *Drosophila* genes was dramatically simplified. Later Fedoroff applied the concept to maize transposons, and these methods are now used worldwide. Thus the Carnegie emphasis on the development of new tools for the advancement of science extends to biology as well as to the physical sciences.

Alfred D. Hershey (1908–1997)

When Milislav Demerec became director of the Department of Genetics at Cold Spring Harbor, he was determined to make the department a leading center for genetics research. One approach he took was to appoint exceptional individuals like McClintock as staff members. Another approach was to sponsor outstanding annual symposia and advanced summer courses at the beautiful and peaceful site on Long Island's North Shore. Perhaps the most important of the summer meetings were those that brought together a small group of scientists working on bacteriophage (bacteria-eaters), the viruses that infect bacteria and are nicknamed "phage." Every known bacterium is preyed upon by at least one type of phage.

Three scientists were the nucleus of what came to be called the "phage group": Max Delbrück, Salvador Luria, and Alfred Hershey. Hershey was born and educated in Michigan while Luria and Delbrück were refugees from a Europe increasingly dominated by Adolf Hitler. At the 1946 Cold Spring Harbor Symposium, Delbrück and Hershey reported experiments showing that phage have their own genes. This discovery and the spread of the phage group's methods and concepts, largely through the summer courses, spawned a whole new field of research that was central to the conversion of genetics to a molecular science. By 1950, Demerec had convinced Alfred Hershey to become a Carnegie staff member and to move permanently to Cold Spring Harbor from Washington University.

The phage that Hershey studied consisted of a core of DNA surrounded by a shell made of proteins. They are easy to maintain in a laboratory, and their

Alfred Hershey in the
laboratory, 1960.

action on their bacterial host cells is striking. Within a few minutes of being infected, the bacterium breaks open and releases a hundred or more identical copies of the original virus. Clearly, something in the virus transmits genetic information to those offspring, but is it the proteins or the DNA? That was the question that Hershey and his colleague, Martha Chase, hoped to resolve.

What was eventually called DNA was first recognized as a component of cells in 1869 by the Swiss chemist Johann Friedrich Miescher. Advances in understanding its chemistry and biology were made over the next 50 years. In 1914 the German chemist Robert Feulgen discovered that DNA would take up a particular red dye, but he considered this result so unimportant that he didn't publish it for 10 years. Later, however, so-called Feulgen staining was used to confirm the fact that DNA appears in all cells and is characteristically located in the chromosomes, which had been so named because they took up a variety of dyes. In the 1920s, the Russian-born American biochemist P. A. T. Levene analyzed DNA and established that its basic building blocks are the four nucleotides: linked units each containing a phosphate group, a sugar, and one of four types of molecules known as bases and abbreviated A, T, G, and C.

Levene correctly deduced that the DNA molecule is built from nucleotide units, but because the proportion of the four bases appeared to be equal in his samples, he concluded that no genetic information could be coded in the molecule. It wasn't until 1950 that Erwin Chargaff at Columbia University showed that the proportion of the bases in fact varied from one species to the next, so that, in principle, DNA could serve as a carrier of genetic information. Chargaff's work also established that particular pairs of bases always occur in equivalent amounts—A equals T and G equals C. This information was crucial to James Watson's and Francis Crick's proposal for the structure of the double helix.

In 1944, Oswald T. Avery, at the Rockefeller Institute, demonstrated to his own satisfaction that the genes of the pneumococcus bacterium were made of DNA. But many biologists remained skeptical and the decades-long debate over whether hereditary information resides in the proteins or DNA of chromosomes continued. Enter Alfred Hershey, master of the experiment.

The 1952 Hershey–Chase experiment was fairly straightforward. Hershey and Chase prepared two batches of viruses. One was grown in a medium stocked with radioactive sulfur-35, which is incorporated into proteins but not into DNA. The other batch was grown in a medium stocked with phosphorus-32, which enters mainly DNA with little entering proteins. Since bacteria are easily infected with viruses, the twin radioactive tracers provided a convenient way to differentiate the fates of the two kinds of molecules.

But infection was only the first part of the process. The viral shells had to be removed from the bacteria for analysis without causing damage. This step is where the blender came in. Named after Fred Waring, the bandmaster, the blender provided the perfect sort of agitation for the purpose at hand. Hershey called his tool an "elegant little cocktail machine." And, as far as the science was concerned, as Hershey put it, "It worked right away."

When the scientists analyzed the results, the implications were clear. The phosphorus-32-rich DNA had gone into the bacteria and the infection proceeded normally. The sulfur-35-rich protein, on the other hand, remained outside; the proteins were apparently unnecessary to the process of viral replication. Thus it was clear that the DNA, not the protein, was responsible for the infection and for the production of new phage.

This famous experiment established clearly that genes are located on molecules of DNA, a central tenet of modern science. The discovery set the stage for the transforming genetic advances of the late twentieth century. In 1969 Hershey shared a Nobel Prize with Max Delbrück and Salvador Luria.

EMBRYOLOGY

The fertilized egg presents one of the central mysteries of biology. How can a single cell carry all the information required to sculpt a fly, a frog, or a human? This question fascinated Franklin Paine Mall, whose driving impulse to understand human development resulted in the establishment of the Department of Embryology in Baltimore.

Franklin Paine Mall (1862–1917)

Franklin Paine Mall was born in Belle Plaine, Iowa, on September 28, 1862. After graduating from the Medical Department of the University of Michigan in 1883, he traveled to Germany where he encountered two extraordinary teachers who changed his life: Wilhelm His, Professor of Anatomy at Leipzig and the most celebrated embryologist of the day, and Carl Ludwig, who is described by Mall's biographer as the "greatest teacher of physiology that ever lived." From these intellectual giants, Mall learned sophisticated experimental techniques for the study of embryology but, more importantly, he developed a life-long commitment to research that was free from the requirements of immediate application.

Franklin Paine Mall by
Thomas C. Corner, from
a photograph taken by
Frederick L. Gates, 1913.

Mall returned to the United States in 1886 and accepted a succession of teaching positions at American universities, culminating in an appointment at the newly established medical school at the Johns Hopkins University in 1893. Over the years Mall acquired a reputation as a gifted teacher. But the dream of establishing a research institute that was devoted to the study of embryology was never far from his mind, and it is in this arena that Mall made his most lasting contribution.

At His's urging, Mall had started a collection of human embryos from autopsies and spontaneous abortions in order to study the changes that occur during gestation. By 1914 he had collected 965 specimens—far too many for a single person to prepare, let alone to study. Mall had long been on the lookout for funding for his project, and the Carnegie Institution now seemed a good source. In his application papers, Mall emphasized an important point: "It may appear that the above outline emphasizes unduly the making of a collection, but this is not the intention, for in my opinion the chief function of an institute of human embryology should be the formation and solution of problems." Some of the problems that Mall hoped to explore through his work were curves of growth, stages of growth in normal development, and pathological changes in embryos. The Carnegie Institution agreed that his goals were worthy, and in 1914 the Department of Embryology for innovative research in development opened in Baltimore on the Johns Hopkins University campus.

Over the next half-century, the Carnegie department would collect over 10,000 embryos. But, as Mall had hoped, the collection proved to have significance that went far beyond the merely archival. Many of the Carnegie specimens were studied in detail. The general technique, which Mall had learned in Germany from Wilhelm His, was to slice the embryos into multiple thin sections. Each section was examined under a microscope to discover detailed information about tissues and organs. Then the outlines of each section were traced and reproduced in two dimensions as well as in plaster discs. The discs were later reassembled, layer by layer, to produce three-dimensional models. Once a large number of embryos had been studied and reconstructed, the models could be placed in a sequence corresponding to developmental stages. The result was a series of "Carnegie Stages," which range from 1 to 23. These stages remain the standard worldwide. The final Carnegie stage represents an embryo at about eight weeks of development, the time by which most organs and tissues have been formed.

Over a period of 50 years, some of the most distinguished human anatomists in the world were associated with the Carnegie department and its collection. John Roc and Arthur Hertig discovered the first two-cell human embryo. Carl Hartman elucidated the menstrual cycle in monkeys. And Elizabeth Ramsey began her distinguished career with the discovery of a 14-day-old embryo.

Elizabeth Ramsey (1906–1993)

Timing is everything in the game of life—from the lives of cells to the lives of humans. This was certainly the case for Elizabeth Ramsey, who, in 1932, while performing a routine autopsy at New Haven Hospital, discovered an apparently normal human embryo that was just 14 days old. One of the youngest embryos known at the time, it was so valuable that everyone agreed that it belonged in the world's finest repository, the Carnegie Collection. The embryo, along with Elizabeth Ramsey, arrived at the Department of Embryology in 1934. Ramsey liked to joke that she was admitted to the department because she "came" with the Yale embryo.

Once she arrived in Baltimore, Ramsey began to teach herself embryology. Her first task was to study the Yale embryo as well as other specimens in the collection. In time, Ramsey learned enough about these embryos to become curator of the Carnegie Collection. By that time, however, she had embarked on a new field of inquiry, thanks to a timely suggestion by George Streeter, director of the department.

Streeter suggested that Ramsey study cells that drifted through the walls of the uterus. This idea opened an entirely new world of discovery for Ramsey. In time, she became an expert on the interactions between the fetus, the placenta, and the mother. Using radioactive dyes and X rays, Ramsey established similarities between monkey and human systems. By the time she died, Ramsey had become one of the world's most respected authorities on the placenta.

Elizabeth Ramsey, ca. 1960.

Later in life, Ramsey called the discovery of the 14-day-old embryo the "most interesting professional thing in my life." Arguably, it was also the most influential "professional thing," since the discovery brought Ramsey to the Carnegie Institution, where a focus on research and a flexible administrative structure allowed her to pursue a long and fruitful career.

Genes and Development

The "Yale" embryo was so small that, at first glance, no one knew what it was. Today, staff members at the Department of Embryology study life on an even smaller scale. This shift in focus occurred during the 1950s, when James Ebert became director of the department. Ebert recognized that advances in biochemistry and genetics opened a path to understanding embryonic development at the molecular level. If the department was to take advantage of this opportunity, geneticists and biochemists, rather than people trained in embryology, needed to be recruited. Also, a new laboratory building appropriate for new directions would be needed.

In 1960 the Department of Embryology moved from its quarters at the Johns Hopkins Medical School to a new building on a wooded site on the university's

Homewood campus. Eventually, the Carnegie Embryo Collection found a permanent home at the National Museum of Health and Medicine in Washington, D.C., where it is available for study. Images of the mounted embryo slices are being digitized. Meantime, members of the Department of Embryology began to explore how genes affect development.

Donald D. Brown was one of the gifted young scientists that Ebert brought to the department. By the time Brown graduated from medical school, he knew that research was what he wanted to do. Inspired by a year at the Pasteur Institute in Paris, where extraordinary work concerning how bacteria regulate the activity of individual genes was in progress, he joined the Department of Embryology in 1961. His dream was to understand the role that genes play in the development of a complex organism from a single fertilized egg cell. Thomas Hunt Morgan had expressed similar thoughts in 1934 in a book entitled *Embryology and Genetics*. However, as Boris Ephrussi, then a visitor at the California Institute of Technology laboratories commented to Morgan, the book did not in fact bridge the gap between embryology and genetics as the title promised. At the time, neither Morgan nor Ephrussi nor anyone else could devise an experimental approach to the role of genes in embryonic development.

Osborne O. Heard, a sculptor, joined the staff in 1913.

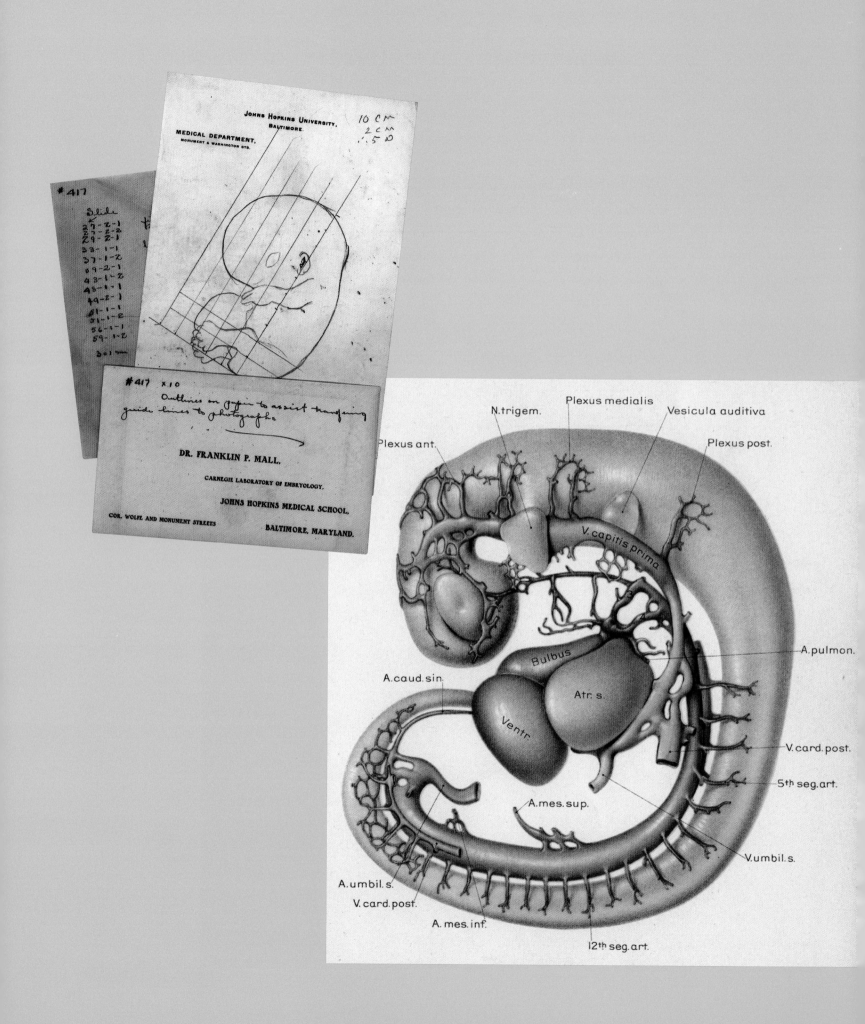

Thirty years later, when Brown arrived in Baltimore, the situation had dramatically changed. By that time everyone agreed that genes are made of DNA and that much of the sequence of DNA nucleotides embodies a genetic code for determining the structures of proteins. Other genes in DNA are copied into RNA molecules that are essential for the manufacture of proteins. Among these RNA molecules are those that enter ribosome particles, the intracellular factories that construct proteins. The ribosomal RNAs are abundant in cells, and DNA carries many copies of their genes. Brown and his colleague Igor Dawid decided to study ribosomal RNA in the frog, *Xenopus laevis.*

Xenopus was a good choice for several reasons. Much was known about amphibian embryology, and the animals were easy to grow in inexpensive bathtubs. Even more important, Brown could obtain mutant tadpoles whose cells lacked the cellular site of ribosomal RNA synthesis, the nucleolus, a dark round body that is easy to see in the microscope. As Brown and his British colleague John Gurdon learned, these mutant tadpoles do not make ribosomal RNA at all because they lack the necessary genes. Normally, the many copies of those genes are located together in particular sites on chromosomes. Those sites are in fact the nucleolar organizer regions that McClintock had identified in chromosomes 30 years earlier. Years before it became possible to isolate genes routinely by molecular cloning, Brown succeeded in isolating the ribosomal RNA genes; they were the first genes to be purified from the cells of complex organisms.

As with any experimental system, amphibians had a limitation as an experimental tool. Their embryology was well known, but their genetics was not, nor are they convenient animals for genetic analysis. Today, the role of genes in embryonic development is studied by Allan Spradling, a *Drosophila* geneticist, who succeeded Brown as department director. Modern methods, many of which were developed by Spradling and his colleagues, make it possible to study the embryology of *Drosophila* in detail. The DNA sequence of the entire *Drosophila* genome is now known, thanks to a cooperative, international effort in which Spradling played a leading role. And because many *Drosophila* genes are very similar to mammalian genes, whatever is learned about flies helps us understand the development of humans.

We are still far from knowing how a human being develops from a single fertilized egg, but thanks to research on frogs and flies, scientists are beginning to glimpse some of the underlying principles. These exciting new developments would no doubt have pleased Franklin Mall who, shortly before he died, inquired of a colleague, "What would you say has been the effect of the Carnegie Institution of Embryology on the research of the [Johns Hopkins Medical School] laboratory?" The colleague replied, "It has lifted the research . . . from a somewhat amateurish to a professional state," to which Mall is reported to have said, "It is exactly what I wished to do."

PHOTOSYNTHESIS

The color of life on our planet is green—green from the chlorophyll in the plants that form the basis for all life, green from the molecules that convert the energy of sunlight into the raw materials of living things. It is little short of astonishing, then, that progress in understanding the basic process by which this happens, the process of photosynthesis, followed such a leisurely pace over the centuries. Human beings understood the movements of planets and stars long before they had even an inkling about the inner workings of the grass beneath their feet.

The first serious study of plant growth mechanisms was undertaken by the Flemish aristocrat, Johannes Baptista van Helmont (1579–1644), who weighed the dirt in a pot, then planted a tree in it. He watered the tree for several years, then weighed the tree and the dirt again. He found that the tree had gained 164 pounds while the soil had lost only a few ounces. Although he came to an incorrect conclusion (he argued that the extra weight came from water), he had established the fact that the bulk of what we call biomass comes not from the soil but from somewhere else.

The Power of the Sun

The role of light was highlighted in the 1779 publication of Jan Ingen-Housz's *Experiments upon Vegetables*. Ingen-Housz (1730–1799) was a Dutch physician who saw that plants absorbed air and exhaled oxygen. He also noticed that this activity decreased in the shade and ceased entirely night. Ingen-Housz inferred correctly that animals and plants are dependent on one another for sustenance and that light plays the critical role in maintaining the bond. But it took another 25 years before the Swiss scientist Nicolas-Théodore de Saussure (1767–1845) added the last ingredient: water. Only then did the final, deceptively simple equation for the process known as photosynthesis emerge: $6\ CO_2 + 6\ H_2O +$ energy from sunlight $\rightarrow$ organic molecule $+ 6\ O_2$. This closed cycle remains the basis on which our understanding of photosynthesis rests.

During the nineteenth century, scientists discovered some of the basic parameters of this process. Chlorophyll was isolated and named in 1818. So were accessory pigments, such as the yellow carotenoids. But it wasn't until the development of the law of conservation of energy in 1845 that the big picture of plant physiology came into focus. For the first time, scientists realized that energy in sunlight absorbed by plants had to be stored somewhere. At this point, the process of photosynthesis became a focus of research. Botanists measured the dependence of the rate of photosynthesis on external factors, such as temperature, humidity, and the amount of sunlight. They also discovered several variants of the chlorophyll molecule.

Thus, by the time the Carnegie Institution was founded in the early twentieth century, some of the preliminary work on the subject had been carried out. But the hard part—understanding how the process works at the molecular level—remained to be done. Plant physiologist Herman A. Spoehr brought this important question to the Carnegie Institution.

Herman Spoehr and the "Mother Reaction"

Herman Spoehr was born in Chicago to a family involved in manufacturing. After receiving his B.S. from the University of Chicago and spending a year in Paris working at one of the leading carbohydrate chemistry laboratories of the time, he returned to his alma mater and received his Ph.D. in 1909. His primary

interest was sugar chemistry, a focus that some scholars have linked to his family's involvement in the candy business. In any case, Spoehr's attention soon turned to photosynthesis, nature's primary producer of sugar. During a two-hour meeting with Daniel MacDougal, director of the Carnegie Institution's Department of Botanical Research, Spoehr made such a favorable impression that he was offered a position as head of the new laboratory devoted to plant chemistry. In 1910, with his new bride, he traveled to Tucson to begin his work.

Herman Spoehr experimenting in laboratory.

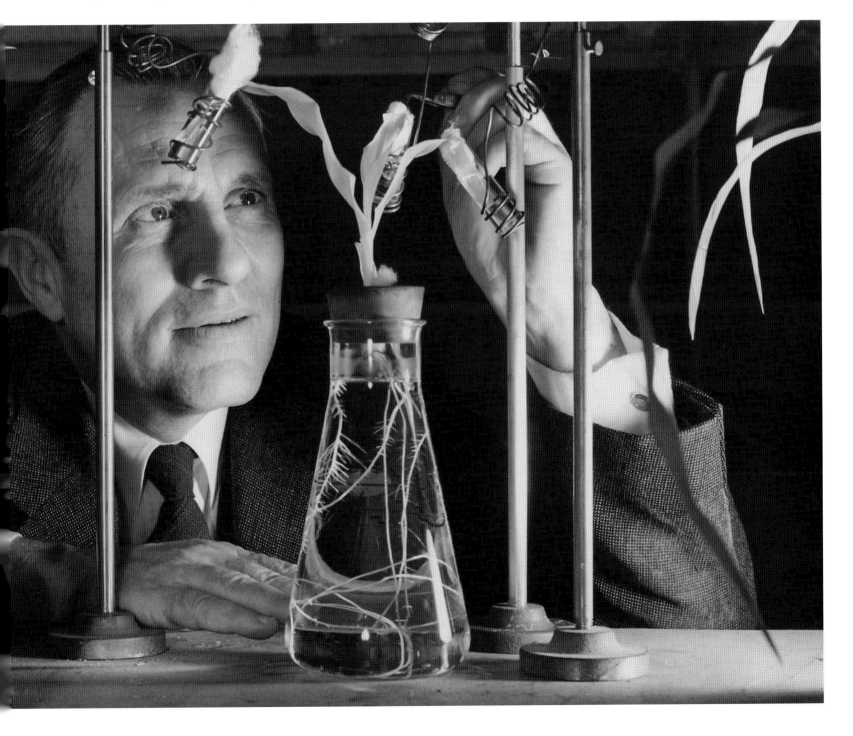

With a few short absences, he remained on the staff of the Carnegie Institution until his retirement in 1950.

When Spoehr took up his position in Tucson, the study of photosynthesis was exclusively a European enterprise. Laboratories in England and Europe, particularly Germany, dominated the field, whereas photosynthesis research in America was in a primitive state. Spoehr was the first American to pursue research in photosynthesis on an ongoing basis, and the research group he assembled remained one of only two in the United States devoted to the subject until the mid-1930s.

Spoehr's hope was to elucidate the roles of a plant's many molecules in metabolism, and he wanted to understand the chemical reactions of photosynthesis, the primary process by which the sun's energy is stored in chemical bonds. Spoehr was fully aware of the enormous importance of photosynthesis. He called it the "mother reaction" of life. He knew that it provides not only all of our food (either directly or indirectly), but that its oil and gas derivatives, buried in the earth for millions of years, also provide most of our fuel. He believed that a scientific understanding of the subject was of prime importance in guaranteeing the world's continued supply of these resources. Though his approach was thoroughly scientific, all of Spoehr's work seems to have been sincerely imbued with this practical understanding. (His keen sense of responsibility, in fact, would eventually lead him, after World War II, to the cultivation of chemically modified algae as an important source of food.)

But where should he begin?

When Spoehr arrived at the Desert Laboratory, he must have been overwhelmed by the possibilities. He began by examining the photosynthetic commodity most in evidence around him: sunlight. He took measurements of daily light intensities at the laboratory, for example, and found that blue light was more intense there than it was on nearby Mount Lemmon, an elevation several thousand feet higher than that of the Desert Laboratory. He also determined that the intensity of the total sunlight was not as important as the intensity of the light's component wavelengths.

Spoehr's research then turned to the study of the breakdown of organic molecules by light in an effort to test some of the current theories about carbon dioxide reduction in sunlight. He quickly found that photosynthesis was much more complex than many previous researchers had supposed. Like most chemists at the time, Spoehr suspected that sugars are the first products of photosynthesis. In which order they were manufactured, however, was not known, nor was their fate after production. Whatever the steps involved, sugars are the only visible products of photosynthesis and thus they provide the only possible glimpse into the plant's complex metabolic activity. It is hardly surprising that Spoehr's interest in photosynthesis would start there. Spoehr didn't know (and wouldn't know until well after World War II) which sugar was first and in what order the rest were produced.

In 1916, Spoehr began a series of experiments in which he sampled tissue from living cacti and took them into the laboratory for analysis. The dependence of molecular concentrations on season, water content, and temperature were measured. Similar detailed experiments established the basic parameters of the molecular processes associated with photosynthesis. The book that Spoehr wrote about these experiments on cacti photosynthesis, *The Carbohydrate Economy of the Cacti*, became a classic.

But Spoehr was never interested in cacti metabolism for its own sake. He was more interested in general principles of photosynthesis, and so he turned to the study of sunflower and bean leaves. That plants occasionally respire as people do—that is, that they emit carbon dioxide and take in oxygen—had been known for many years. Exactly how this reaction occurred, however, and how it was related to photosynthesis was not yet understood. It was also not known to what extent photosynthesis functioned independently of a plant's general metabolic activity. Spoehr reasoned that if photosynthesis and respiration were related, it would follow that the factors that affect respiration would also affect photosynthesis.

This was indeed the case. There exists a "delicate adjustment" between the two, Spoehr concluded in 1920. When respiration is low, so is the photosynthetic rate. Furthermore, when a plant is deprived totally of oxygen, photosynthetic activity ceases entirely, even in the presence of bright sunlight. It was hard to figure what was going on. How was it that photosynthesis required oxygen when oxygen was also its end product? This riddle would puzzle Spoehr and other scientists for many years.

Greener Pastures

With his research moving away from desert plants, there was no reason for Spoehr to remain at the Desert Laboratory. Several of the Institution's plant ecologists had already shifted their research to the more temperate climate of Carmel,

C. Stacy French (left) and Harold Milner in laboratory at Department of Plant Biology, 1960.

California, and in 1921, the Institution began construction of a new chemical laboratory in Carmel for Spoehr. By August the building was ready for occupancy and Spoehr's apparatus was shipped from Tucson to Carmel. Clearly, the Institution was committed to Spoehr and his work on the sugars. Likewise, Spoehr remained committed to the Institution. As he became better known, he received a continuing stream of offers from other research establishments and universities. Although the money offered was often more than Carnegie could afford, Spoehr turned these offers down—all but one, which was a temporary interlude.

Both parties benefited from this long-term relationship. By the mid-1920s, Spoehr had become the nation's leading authority on photosynthesis, and his book *Photosynthesis* (published in 1926) remained the standard reference on the subject for over a decade. Not surprisingly, supporting the leader of such an important field of inquiry was gratifying to the Institution's president, John C. Merriam. For his part, Spoehr appreciated the Institution's laissez-faire approach to science funding—especially after 1923, which is approximately when Spoehr admitted openly that photosynthesis was not the simple process that he and many others had hoped it would be. After a decade and a half of study, it was clear that photosynthesis involved numerous coupled chemical reactions. It would take an ongoing, noninvasive support system such as the Carnegie Institution provided to unravel the complexity.

Department of Plant Biology, Stanford.

In Spoehr's view, the only way to make progress in the field was to approach the problem as an experimental one rather than trying to develop theories based on partial and incomplete information. The validity of this approach was recognized by the Institution with Spoehr's appointment in 1928 as chairman of the newly formed Division of Plant Biology (later the Department of Plant Biology), centered in yet another new laboratory, this one on the campus of Stanford University. Here Spoehr's experimental program was put into action. The main objective of this new facility was to pursue the chemical details of plant physiology, particularly photosynthesis, through the study of biochemical reactions.

During the 1930s, the division made a series of fundamental advances. Its scientists worked on the role of yellow leaf pigments like carotenes in photosynthesis, working out the complex chemical structures of these molecules and the method by which they absorb light. Division researchers also made discoveries regarding the amount of energy needed to process a molecule of carbon dioxide, one of the most fundamental pieces of information about the process of photosynthesis. Working with instrument makers at Mount Wilson, scientists at the division built instruments capable of generating and measuring light in photosynthetic reactions. They used this apparatus to show that the energy required by the process was much greater than anyone had suspected, thereby demonstrating that the number of steps involved in photosynthesis was greater than had been thought. And the division pioneered new techniques, including the use of radioactive tracers in following atoms of carbon through the photosynthetic process, the application of spectroscopy to studies of various photosynthetic pigments, and procedures for isolating chloroplasts, the components of cells that carry on photosynthesis.

Many details of photosynthesis are still unknown, and to this day, the process of converting sunlight into chemical energy remains a focus of Carnegie research. But thanks to Herman Spoehr and his successors, we are significantly closer to understanding the chemistry of life on our green planet.

TRANSCENDING BOUNDARIES

ECOLOGY

*A*t the end of the nineteenth century, an accident of geography presented scientists in the United States with a unique opportunity. Unlike European countries, which lie entirely in Earth's temperate regions, the United States includes a part of the great desert area that stretches southward from Arizona and New Mexico. With the establishment of a desert laboratory by the Carnegie Institution in 1903, American scientists seized a unique opportunity to study the adaptations of plants to desert conditions and to learn about desert ecosystems. Located on a hill two miles west of downtown Tucson, Arizona, the laboratory was the first botanical facility in the world dedicated to the study of desert ecosystems.

Ecology is the study of the complex relationships among plants, animals, and their physical environments. Its practitioners are concerned with the distribution of organisms, the competition among them, and the causal factors underlying their local adaptations. As such, ecologists have to adopt a broadly interdisciplinary approach to the study of life. An understanding of local rocks, soils, weather, and climate must be integrated with studies of organisms at scales from individual molecules to broad geographical habitats.

Desert Laboratory ecologist Forrest Shreve with tripod.

The decades around the turn of the century were pivotal to the development not just of ecology but to all of biology. It was during these years that the field began to splinter into distinctive disciplines, mirrored by the concurrent formation of professional societies. Ecology was an offshoot of botany, which itself was breaking away from the traditional emphasis on natural history to subdisciplines with more rigorous attention to experimentation. It was within this changing scientific climate that the Carnegie Institution seized the opportunity to establish a laboratory in the desert.

Extremophiles

At the turn of the twentieth century, the cattle town of Tucson was as rough and raw as any other young American settlement. Though it boasted a university and an agricultural experiment station, it was also a place where gentlemen carried pistols. It was into this hot and dusty outpost that two young botanists stepped one day in January 1903. Daniel MacDougal and Frederick Coville, both established scientists in their mid-thirties, were no strangers to the desert. Coville had been botanist for the U.S. Department of Agriculture (USDA) expedition to Death Valley in 1891 and nine years later had made a research trip across Arizona. MacDougal had been a member of another 1891–1892 USDA expedition to Arizona and Idaho and a few years later had undertaken a plant-collecting trip into Navaho territory.

Though their 1903 journey was exploratory, MacDougal and Coville were investigating real estate rather than plants. The newly established Carnegie Institution had sent them to find a suitable place to build a desert botanical laboratory. In 1902, Carnegie trustees pledged $8,000 for construction and one year's operation of such a facility. The ideal site, according to the Institution's Advisory Committee in Botany, would provide a desert climate, flora as plentiful and varied as possible, easy accessibility, and some degree of habitability.

Tucson met all of these criteria. Furthermore, Tucson's Chamber of Commerce offered subsidies for 40 acres of land on Tumamoc Hill, a gentle and undisturbed rise about two miles outside the town. Later, the Chamber of Commerce agreed to install, free of charge, water, telephone, light, power, and a road.

Approval came in March and progress was rapid. Once the road was finished and the architect's design approved, construction began. By September 1903, the Desert Laboratory was ready to begin its work.

In the beginning, the trustees voted to fund the laboratory for a five-year trial run. Ecology was an infant science, and Arizona was a long way from the civilizing influences of a big city. In this respect, the Desert Laboratory was considered an experiment. Even its organization was experimental, for at the outset it had no director, no "exceptional man" to guide the research.

Getting Started

Surprisingly, the first investigator to take up residence at the Desert Botanical Laboratory was not an ecologist, nor was he especially interested in desert plants. William Cannon, a plant anatomist, had been employed as an assistant at the New York Botanical Garden. Cannon had just received his Ph.D. when he was offered the Carnegie position.

His task was formidable. Starting work with little more than an invitation to be useful, he held the future of the laboratory in his hands. His meager instructions were to inquire into the "morphology, physiology, habit, and general life history of the species indigenous to the deserts of North America." This mission must have been heady stuff for a young botanist most recently confined to a laboratory and now surrounded by miles of uninterrupted desert.

When Cannon arrived in August 1903, only the town and the rugged mountains rimming the valley provided relief from the scrubby desert landscape. Dominating the valley scene were the creosote bushes, interspersed with several species of prickly pears and joint pine. On the foothills were armies of the familiar saguaro cactus, which was named *Carnegiea gigantea* in honor of Andrew Carnegie.

Many of the plants of the Sonoran Desert were represented on the grounds surrounding the laboratory. Before deciding which ones to investigate, Cannon focused on completing the building. By November, with construction complete and his wife and child established in a tent at the base of the hill, Cannon was ready for work. His rigorous studies of cactus anatomy and physiology set the tone for subsequent efforts.

By 1905, at the recommendation of Carnegie president Robert Woodward, the desert laboratory had become a bona fide research department of the Carnegie Institution. While unofficially it would continue to be called "the Desert Laboratory," the name gracing its new letterhead was more specific: the

Department of Botanical Research. Departmental status raised the obvious need for a director, and President Woodward offered the job to Daniel MacDougal. MacDougal, a plant physiologist with expertise in desert flora, was lured from his job as assistant director of the New York Botanical Garden.

With an increase in the department's annual budget from $6,000 in 1904 to $33,000 in 1906, MacDougal added scientific and support staff. He also oversaw improvements to the physical facilities. The enlarged grounds were fenced, and the laboratory building was doubled in size to include a greenhouse and three additional rooms equipped for studies in physiology.

Plants and Water

At first, the laboratory's research focused on water, the critical limiting factor for desert plants. An understanding of how plants use water became a central problem on the agenda, but it was not toward the practical science of agriculture that the Desert Laboratory pioneers directed their work. Rather, they sought an ecological understanding of plant adaptation and distribution, an understanding that would inform the principles that lay at the very foundation of agriculture and forestry practices.

Plants use the simple mechanism of water tension for pumping water from roots into stems and leaves. If more water is released, or transpired, from the leaves, then more water must be pulled up from the roots. Laboratory research soon revealed that succulent desert plants have extraordinary abilities to take up and store water. A barrel cactus contains almost 80 percent water. One of these cacti, a victim of scientific curiosity, was kept in a waterless (but still living) state in the laboratory for six years. It lost only 5 grams a day for a total of 24 pounds. The staff affectionately called it Algernon. Once its ordeal was over, Algernon was transplanted to the garden outdoors where it thrived for many years.

While desert plants retain much of the water they absorb through their roots, they also transpire, often at a surprising rate. The amount of water lost from a plant through transpiration is largely determined by the evaporative power of the surrounding air. Measuring this evaporation rate proved difficult, owing to the lack of a suitable instrument. A solution was devised by visiting scientist Burton Livingston, of the University of Chicago, who constructed such a device at the desert lab. His "atmometer" consisted of a hollow clay sphere

Carnegie Institution president Robert Woodward at the Desert Laboratory, 1906: Standing left to right: Mr. Reeder, Godfrey Sykes, William A. Cannon, Burton E. Livingston, Francis E. Lloyd, Mr. Lantz. Seated, left to right: Mrs. Sykes, Charles Davenport (Director, Department of Experimental Evolution), Robert S. Woodward, Daniel T. MacDougal, George Shull, Mrs. Livingston.

fitted over a glass tube. The tube was inserted through a hole into a cork, which was fitted into a jar of water. As moisture evaporated from the sphere's clay surface, additional water was drawn up into the tube from the jar. The water level provided a measure of the air's evaporating power.

Using atmometers, 12 of which were installed on the desert lab grounds, Livingston was able to quantify the relationship between transpiration and evaporation. This discovery allowed him to compare the physiological regulation of transpiration among different types of desert plant species. He also determined that plants are able to regulate their own water loss. Livingston would go on to publish numerous papers as well as a massive treatise, *The Distribution of Vegetation in the United States as Related to Climatic Conditions,* which was coauthored by fellow–Carnegie ecologist Forrest Shreve and published in 1921.

Forrest Shreve

Forrest Shreve, perhaps more than any other scientist, came to be associated with Carnegie's desert laboratory. A modest and quietly methodical man, Shreve was an ecologist with a passionate interest in experimental physiology. As a staff member at the desert laboratory from 1908 until his retirement 40 years later, he developed a love for the desert and an intimate knowledge of its ways. He was, he later wrote, "not an exile from some better place, but . . . a man at home" in the desert.

In 1910, he was joined in Tucson by his new wife, Edith Bellamy Shreve, who was a physicist and chemist. It was rare at that time for a woman to aspire to a scientific career, and rarer still to be paid for the work. Nevertheless, Edith was able to carve out an experimental research program for herself at the desert laboratory, receiving considerable psychological support from her husband (who did most of the cooking) and some monetary support from MacDougal (who gave her funds for equipment). She never received a salary from the Institution, though she didn't complain; she simply went to work on plant transpiration. It was she, for example, who discovered that desert plants lose water at night and take it in during the day—a result that was in direct opposition to the conventional wisdom of the time.

In his first years at the Desert Laboratory, Forrest Shreve, a native of Maryland, concentrated on learning all he could about the desert and its plants.

The desert was an ideal place for him to apply his ecological training and principles, and it provided the opportunity for him to pioneer studies in a little-understood region of the globe.

Shreve's work emphasized the impact of the physical environment on the evolution of plant communities, rather than interspecies competition. In this respect, he anticipated topics in ecology that would be revisited by many scientists in the latter part of the twentieth century. Shreve began his research by exploring transpiration, comparing the process in desert and tropical plants. He learned that desert plants transpired more than those of the tropics, a conclusion that went against conventional wisdom. He also studied the factors underlying the germination of desert seedlings and the related issue of how plant populations become established in the desert. He monitored seed germination of several plants, both in the laboratory and in the field, and he calculated the approximate ages of these

Desert Laboratory director Daniel T. MacDougal (first car) and Edith and Forrest Shreve in western Arizona, 1915.

Edith Shreve on horseback.

Plate from The Cactaceae, *vol. II, by N. L. Britton and J. N. Rose, 1919-1923.*

plants on the grounds of the lab. The resulting curves for growth rates led to the alarming conclusion that young saguaros were not present in sufficient numbers to maintain the population. The reasons for this decline (temporary, as it turned out) would not be understood until many years later.

Shreve's interest in the germination of desert seedlings led him to conclude that physical factors in the environment, not biological ones, were of primary influence. He would come back to this idea again and again as he investigated how plants responded to such physical factors as temperature, rainfall, and soil moisture. He published his conclusions in *Vegetation of a Desert Mountain Range as Conditioned by Climatic Factors*, issued by the Institution in 1915 as publication number 217. However, his book was little noted and even MacDougal was unimpressed. The reason was simple: a new staff scientist, Frederic Clements, had by that time assumed such a dominant role in ecology that he overshadowed everyone else.

PLATE XXXVI

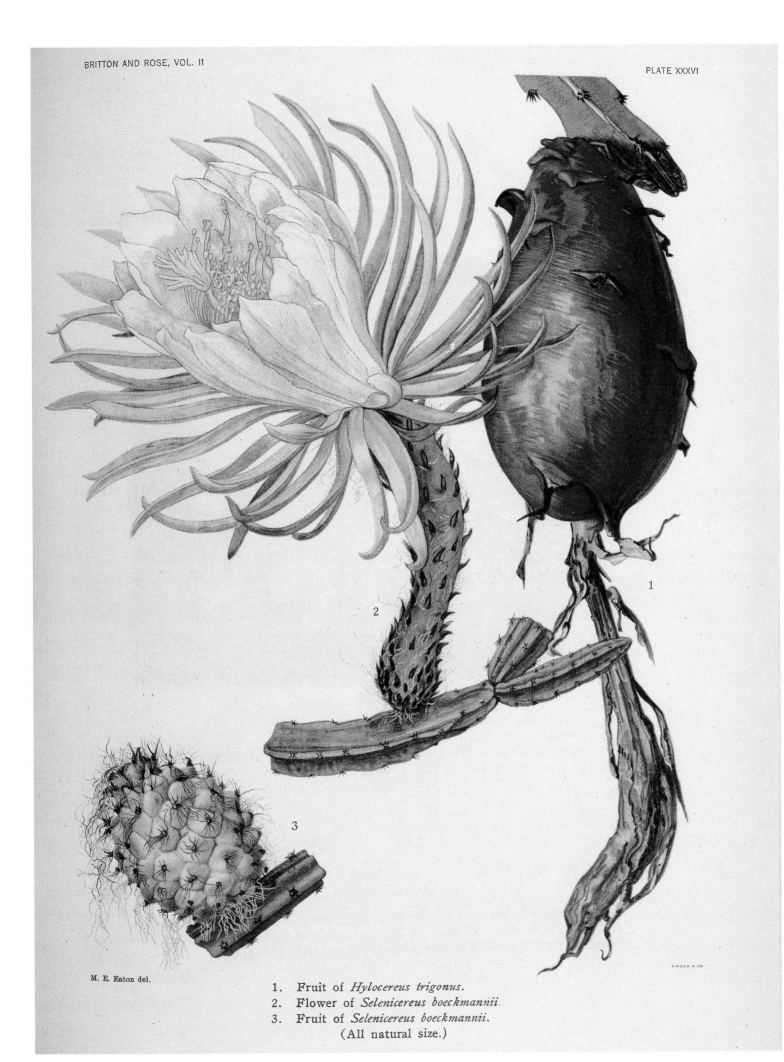

M. E. Eaton del.

1. Fruit of *Hylocereus trigonus*.
2. Flower of *Selenicereus boeckmannii*.
3. Fruit of *Selenicereus boeckmannii*.
(All natural size.)

Frederic Clements

Clements came to the attention of the Carnegie Institution in 1903, when he applied unsuccessfully for a grant. Ten years later, he came into contact with Daniel MacDougal, who found his ideas appealing, especially his rigorous support for ecology as a physiological science. Within a year, Clements was offered a part-time position, and in 1917 he was appointed a permanent associate—a staff member, but without a department. Though he was still loosely connected to the Department of Botanical Research, Clements was given independent funding to establish his own laboratory. Able to devote himself full-time to research, he and his wife split their time between Tucson and Manitou Springs, Colorado, where their Alpine Laboratory was nestled in the mountain forest. This facility became

Frederic Clements observing dunes in the Great Plains during the drought of 1934.

*Uaxactún—Groups A and
B. The published version,
shown here, was produced
from an original watercolor
by Carnegie archeologist,
Tatiana Proskouriakoff.*

Proskouriakoff undertook her most ambitious work after the Carnegie
department had closed. Still a Carnegie research associate, she studied Mayan
hieroglyphics in great detail and came to a startling conclusion. She suggested
that the glyphs were not records of the feats of gods—as was formerly supposed
—but the historical records of the events in the lives of real people. This discovery is considered a major breakthrough. At the time of her death, Proskouriakoff
had nearly completed a history of the Mayan Classical period A.D. 300 to 900.

Anna Shepard (1903–1973)

A Carnegie staff member from 1931 until her retirement in 1968, Anna Shepard pioneered the use of petrographic and chemical analyses of pottery. She persuaded several generations of archeologists that such evidence, when integrated with other findings, could be used to give a more comprehensive picture of the past than would otherwise be available.

Born in New Jersey in 1903, Shepard received her B.A. from the University of Nebraska in 1926. Working at many archeological institutions over the years, she embraced the study of pottery and particularly the technology of glazing as her chosen field. Her technique was to shave off a small sample of pottery glaze, then examine it under a microscope to identify the mineral grains it contained. From this sample, she could tell where the raw materials came from, thereby establishing the existence of trade relationships between distant groups. The technique also provided a way of documenting changes in technology over time. She worked both in the American Southwest and in Mayan country. She is credited with establishing the importance of women as commercial potters in the southwestern economy.

Her celebrated treatise, *Ceramics for the Archaeologist,* published by the Institution in 1954, remains a Carnegie best-seller.

Anna Shepard working at her Boulder, Colorado, laboratory, early 1930s.

BIOPHYSICS

Among the great achievements of nineteenth- and early twentieth-century science was the realization that all life is based on chemistry—that organisms may be viewed as systems of chemical reactions. The old doctrine of vitalism, which held that living things are uniquely imbued with a "life force," had been thoroughly discredited. The romantic notion that something different characterized the complex molecules of living systems—that perhaps even a different set of laws governed their operation—could no longer be supported. From the simple hydrogen atom to the most complex living cell, the laws of chemistry reigned supreme over all matter.

But how does a single cell obtain energy from its surroundings? What are the chemical mechanisms of microbial metabolism? For most of the first half of the twentieth century, these questions, among the most central problems in biochemistry, seemed intractable, for it was difficult to imagine how to design an experiment that would trace the chemical reactions of molecules within a living cell. Then, soon after the end of World War II, detailed answers began to emerge from one of the Carnegie Institution's boldest ventures: the new field of biophysics at the Department of Terrestrial Magnetism.

It would be difficult to name a pair of scientific disciplines less similar in their scope and context than cellular biology and nuclear physics. Prior to World War II, the study of microbes relied in large measure on powerful optical microscopes, selective staining techniques, and qualitative descriptions of cellular behavior. Nuclear physics, by contrast, boasted a two-pronged strategy of exacting experimental measurements coupled with abstract theoretical modeling. The Carnegie biophysics effort, which required expertise in both nuclear physics and cellular biology, was thus the epitome of interdisciplinary research.

Merle Tuve, Lawrence Hafstad, and Odd Dahl with high-voltage apparatus used to obtain gamma- and beta-rays. 1932.

Merle Tuve and the Strong Force

The roots of the Carnegie Institution's biophysics venture can be traced to a quarter-century earlier and the arrival of physicist Merle Antony Tuve at the Department of Terrestrial Magnetism. Born in Canton, North Dakota, in 1901, Tuve received his undergraduate degree from the University of Minnesota in 1922, then went on for graduate training at the Johns Hopkins University. Tuve's first contact with the Carnegie department occurred shortly after he arrived in Baltimore when Gregory Breit, a theoretical physicist whom Tuve had met during his Minnesota days, invited him to collaborate on studies of the ionosphere. In a historic experiment in 1925, the two men used radio-echo sounding equipment to verify the existence of the ionosphere. This achievement paved the way for the development of radar. It also furnished a thesis topic for Tuve, who received his Ph.D. in 1926.

On graduation, Tuve's interest centered on the atomic nucleus, and he was sorely tempted to travel to Ernest Rutherford's laboratory in England. Instead, he was enticed to join the staff at the Department of Terrestrial Magnetism, where he remained until he retired. During this extraordinary 43-year career, Tuve displayed a remarkable ability to identify and pursue fruitful avenues of inquiry for the times. In the early 1940s, he oversaw the development of the proximity fuse, a device that contributed greatly to the victory of the Allies. As director of the department from 1946 to 1967, he instituted research programs in seismology, radio astronomy, and geophysics. He was a key player in the development of image tubes, which transformed observational astronomy during the 1960s and 1970s. But in 1926 Tuve's mind was on nuclear physics. One of his most ambitious undertakings was an experimental program to understand the force that holds the particles in an atomic nucleus together.

The atomic-physics observatory at the Department of Terrestrial Magnetism, October 20, 1938.

In tackling this problem Tuve and his colleagues were following a tradition of more than three centuries of physics research. In the mid-1660s, while on leave from plague-ridden Cambridge University, Isaac Newton formulated his famous laws of motion. The first of these great laws includes the definition of a force as a phenomenon that causes a mass to accelerate. Newton formulated a description of the force of gravity. Other investigators contributed to a unified view of electric and magnetic phenomena—the electromagnetic force—which is the other obvious force in our lives.

Discoveries about the structure and behavior of atoms ultimately led to the recognition of two other, more subtle forces that operate only at the extremely short distances of the atomic nucleus. The so-called weak force plays a role in certain types of radioactive decay. The other nuclear-scale force, called the "strong force," overcomes electrostatic repulsion to bind positively charged protons. In February 1935, the Japanese theorist Hideki Yukawa described the essential characteristics of this force, without which the nucleus would fly apart. But no one had measured the exact magnitude of the strong force. Nor was it known whether the force in proton–proton interactions was the same as the force in proton–neutron interactions.

In 1935, Tuve and his colleagues Lawrence R. Hafstad and Norman P. Heydenburg, in collaboration with Gregory Breit, decided to tackle the problem by measuring proton scattering during proton–proton collisions. The experiments were straightforward in principle: simply observe the scattering of a stream of protons that were accelerated by the powerful two-meter Van de Graaff electro-static generator. This landmark machine, with an electric potential of approxi-mately 1.2 million volts, developed more than enough proton acceleration for the task. (In fact, such high electric potentials led to cascades of X rays—a dangerous phenomenon that led Dr. Winifred Whitman, a scientific collaborator at the Department of Terrestrial Magnetism and the wife of Merle Tuve, to invent the now ubiquitous X-ray film badge in 1930.)

The proton–proton experiment was a challenging undertaking, requiring sig-nificant advances in instrumentation by the Carnegie group. Furthermore, mak-ing any sense of the data required a sophisticated quantum-mechanical analysis, the techniques of which had to be worked out by Breit. The results of theory and experiment showed that the strong force was attractive, even though the particles separated after colliding. Interpretation showed the force to have a very short range compared with the electrostatic force, which was also present. In the 1936 *Year Book* of the Carnegie Institution, Tuve wrote, "An important new physical force, in its way as fundamental and significant as the force of gravita-tion and that of electric attraction and repulsion, was directly observed and measured for the first time in the Department's laboratory this year."

This groundbreaking experiment was followed by another experiment that recorded the scattering of neutrons by protons. To succeed in this new experi-ment, a completely different set of technical problems had to be overcome. Results depended on the tedious observation of thousands of cloud chamber photographs, which revealed protons recoiling from collisions with neutrons in the hydrogen-filled chamber. The analysis showed the proton–neutron force to be identical with the proton–proton force. This equivalence immediately formed the basis for calculating models of nuclear structure, and it remains so today.

War and Transition

Buoyed by their success and armed with a state-of-the-art accelerator laboratory, the Department of Terrestrial Magnetism's physics program expanded. Studies of proton scattering were complemented by work on neutron absorption and the production of isotopes, atoms with controlled numbers of neutrons. Tuve even touched on the field of biophysics in his collaborative studies with scientists at the Department of Genetics on mutagenic effects of neutrons on *Drosophila*.

The Carnegie nuclear physics program expanded in 1939 with the arrival of Philip H. Abelson, who was to become one of the most influential figures in the Institution's history. Abelson was born in Tacoma, Washington, in 1913, and received his B.S. and M.S. degrees in chemistry at Washington State University. Moving to the University of California at Berkeley, he did his doctoral research in the new field of radiochemistry as a member of Ernest O. Lawrence's distinguished research team. Abelson carried out one of the first studies of uranium fission products, and he is the co-discoverer of the element neptunium.

Abelson received his Ph.D. from Berkeley in 1939 and immediately joined the staff of the Department of Terrestrial Magnetism, where he participated in the design of a cyclotron like the one invented by Ernest Lawrence. A cyclotron is a machine that accelerates charged particles like protons and then arranges for the particles to strike a target.

When high-energy protons strike an atom's nucleus, they can trigger nuclear reactions by breaking up the nucleus. Thus the cyclotron can be used to create radioactive isotopes of chemical elements for use in experiments. These isotopes, because they are unstable, are often rare in nature, but they can be created copiously in the machine. It was precisely this use to which Abelson and his colleagues planned to put the new cyclotron.

World War II forced many scientists to refocus their efforts. Tuve concentrated his energies on the development of the proximity fuse. Abelson turned his attention to the separation of uranium isotopes for the Manhattan Project. The development of the atomic bomb was crucial to the Allied victory, but once the war was won, many nuclear physicists became uncomfortable with the implications of their work. The time was ripe for a new direction in physics research.

Shift in Focus: Biophysics

Returning to the Department of Terrestrial Magnetism after the war, Abelson began to organize a new research team, consisting mainly of physicists, to address fundamental problems of biochemistry with new physical methods. Metabolism, the chemical processes by which cells obtain structure and energy, would be their ambitious central focus. Knowing that a living system is a collection of chemical reactions is a far cry from being able to say what those chemical reactions are. Even the simplest microorganism regulates hundreds of complex chemical reactions. In the first part of the twentieth century, scientists attempted to map out the chemical pathways of life using techniques that were cumbersome and difficult by today's standards. By the end of World War II, many of the molecules involved in basic metabolism had been identified, but the chemical pathways of metabolism were largely unknown.

In 1947 a team of scientists including Richard Roberts, Dean Cowie, Ellis Bolton, Roy Britten, and Philip Abelson decided to use radioactive isotopes as chemical tracers to study these metabolic reactions. Following a pivotal visit by Roberts to the Department of Genetics at Cold Spring Harbor, the group picked a bacterium called *Escherichia coli* for study. This bacterium plays the same role in the study of single-celled organisms that the fruit fly does for genetics: It has become the laboratory model of choice for biochemical experiments. The problem

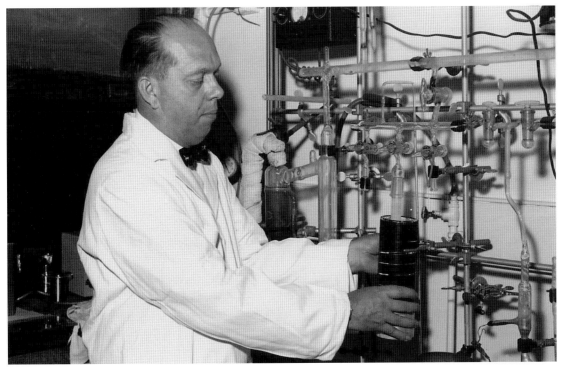

Physicist Philip Abelson prepares a sample for isotope analysis to study how biomolecules are preserved in the fossil record.

that the group set for itself was to work out, in detail, the chemical processes that went on in this bacterium.

The basic design of the experiments, first proposed by Roberts, was simple. Bacterial cultures were grown and fed with materials containing radioactive isotopes of common elements like carbon, phosphorous, sulfur, and sodium. The physicists were well trained in the use of these isotopes and they had at first expected to use a supply from their own cyclotron source. However, the isotopes needed for the group's work were soon available from reliable commercial producers and the cyclotron was shut down.

After several bacterial cell divisions, the bacteria were pulverized and their molecules were separated and analyzed, most often by the technique of liquid chromatography. Liquid chromatography depends on the fact that when a liquid moves, it tends to drag molecules with it. In this method, a collection of molecules is spotted onto one end of a sheet of paper. That end is then placed in a liquid solvent whose properties are known. Ordinary capillary action causes the liquid to rise up in the paper, carrying the lighter, more mobile molecules faster and higher than the heavier, less mobile ones. How fast a molecule moves depends on many factors, including its size and the arrangement of its electrical charges. Consequently, after some time has elapsed, each species of molecule winds up at a different place on the paper. Even a complex assemblage of molecules can be separated into its components with this method.

The key to tracing molecular reactions in *E. coli* was to identify which of the bacterium's new molecules had become radioactive. The most common technique for detecting radioactive decay was simply to lay a photographic emulsion on top of the paper, then look for the dark spots where the radiation had exposed the film.

By 1954, after six years of intensive experimentation, the group had assembled a massive amount of information about the molecular workings of *E. coli*. The book that the five researchers published in 1955, *Studies in the Biosynthesis in Escherichia coli*, marked the first time that so many of the details of the metabolism of an organism had been worked out. The book became a standard handbook, a reference for an entire generation of molecular biologists. As such, it had an enormous influence at a crucial point in the growth of the field.

Abelson went on to a distinguished career that spanned many areas of science. In 1953, he became director of Carnegie's Geophysical Laboratory, and then, from 1971 to 1978, he served as president of the Carnegie Institution.

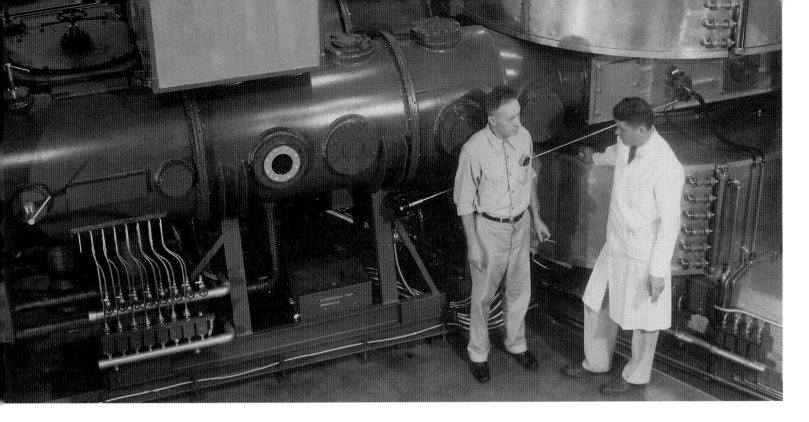

Dean Cowie (right) at the new cyclotron

Since then he has been an institutional trustee. The headquarters building of the Department of Terrestrial Magnetism is named for him.

His association with the Institution has spanned 62 years. At the same time he undertook an astonishing variety of important positions. In 1962, for example, he became editor of the prestigious journal *Science*, the flagship publication of the American Association for the Advancement of Science. At the same time he was an adviser to a dozen important government groups. When he was chided for attempting to do too much, his response was simple: "I have observed," he said, "that most of those who think I am attempting to do too much and will fail in the attempt have only one job and don't do that very well."

Abelson's enormous energy, his ability to achieve, and his conscientious approach to scientific inquiry is unusual, but not unique. The Carnegie Institution has supported many such people over the years. This book has touched on some of them, but there are many more. They focus on different questions and employ a range of methods to find answers. Even so, they share a number of attributes. Philip Abelson's memoir of Merle Tuve contains as good a description as one is likely to find of the "exceptional scientist" whom Andrew Carnegie so fervently sought:

"Tuve was a dreamer and an achiever, but he was more than that. He was a man of conscience and ideals. Throughout his life he remained a scientist whose primary motivation was the search for knowledge but a person whose zeal was tempered by a regard for the aspirations of other humans."

As it begins its second century, the Institution continues to support men and women like this.

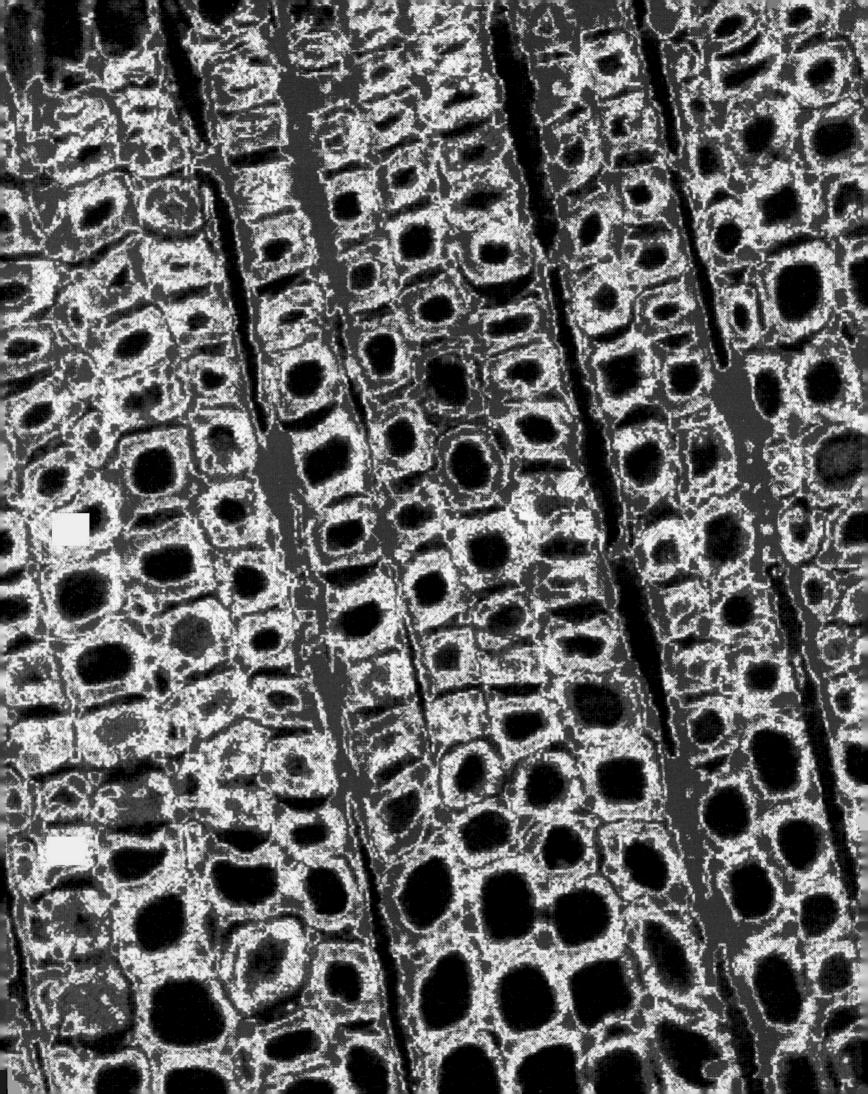

THE ENDLESS FRONTIER

The Carnegie Institution in the
21st Century

The Institution's founders shaped a research agenda that sustained a century of discovery. Yet, they could not predict the paths that would be taken by science in those 100 years or the ways in which Andrew Carnegie's institution for discovery would impact the directions and outcomes. No more can anyone see now where the next century will lead. At best there may be a glimpse of the very near future in the questions that Carnegie scientists now find compelling.

Astronomy

Fossils occasionally preserve the atoms of once-living organisms. The cellular structure of 400-million-year-old wood is revealed in this map of carbon concentration. Red represents the most carbon-rich areas.

The Magellan telescopes at Las Campanas can catch photons from the farthest reaches of space and thus see back in time to the early universe. Complemented by telescopes in space and those that see at wavelengths outside the visible range, they will permit astronomers to learn more about how structures formed and grew in the universe. How did the inhomogeneities in the gas formed soon after the big bang lead eventually to the formation of stars? How were galaxies assembled out of the clumps of gas and stars in the early universe? These daunting questions will stretch the capabilities of current telescopes and instruments.

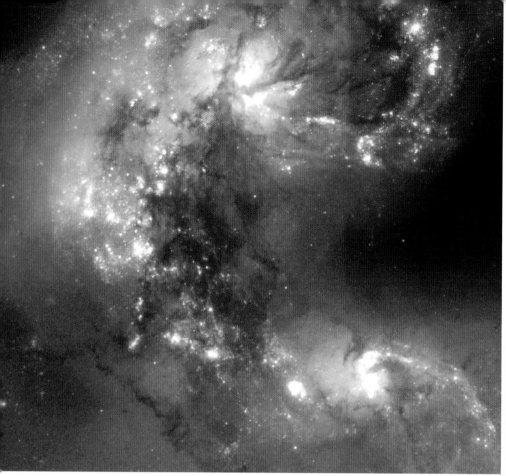

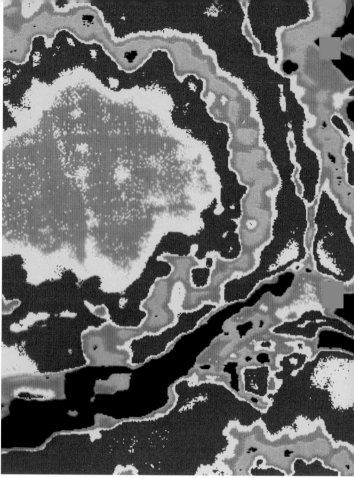

Astronomers are already planning for the next generation of spaced-based and giant, land-based telescopes. Light of all wavelengths is the tool of astronomy, astrophysics, and cosmology.

The questions being pondered by other Carnegie scientists deal with material things, that is, atoms and molecules. There is a profound connection between these pursuits because stars are the ultimate source of the materials on Earth. All but the few lightest of the chemical elements originated in stars, so the history of the elements is embedded in the history of stars. First-generation stars, presumably built predominantly from the hydrogen and helium produced in the big bang, have yet to be detected. Second-generation stars incorporated heavier elements ejected from dying first-generation stars. The study of such stars yields information about the process of element production and the birth of our galaxy.

How are the elements that are ejected from dying stars used to construct new stars and planets? How and from what materials was the solar system constructed? What processes were involved? Carnegie theoreticians are interested in testing competing theoretical models for how planets of Jupiter's mass could have formed in our own solar system. A Carnegie scientist is spearheading Project Messenger, the first NASA mission in 25 years to study planet Mercury. Complementary questions are asked by geochemists and cosmochemists. They have acquired powerful new instruments to obtain clues from meteorites and ancient rocks. Multiple-collector inductively coupled plasma mass spectrometers (MC-ICP-MS)

Left: Pair of colliding disk galaxies (NGC 4038/4039), as imaged from the Hubble Space Telescope. The galaxies' old centers show up in bright orange; the newborn stars and clusters reveal their youth through blue colors.

Right: Carbonate globule from a 4.5 billion-year-old Martian meteorite (ALH 84001). The carbonate is water-bearing and thus a possible indicator of ancient life on Mars.

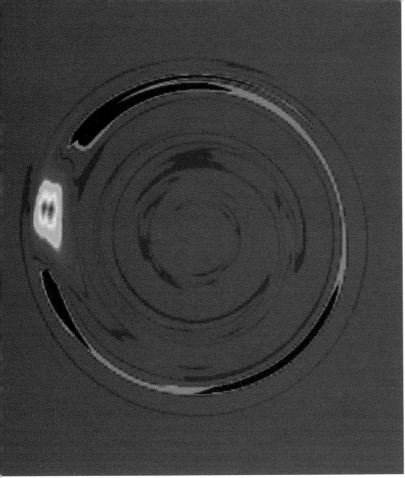

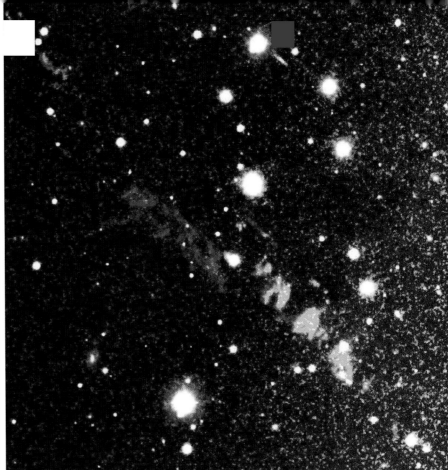

Left: Carnegie theorists create computer models like this one to understand how planetary systems form.

Right: This image was taken during the first week of observations at the 6.5-meter Baade telescope. Most of the bright objects are from our galaxy, but the bright object near the center represents globular clusters in NGC 5128, the closest giant elliptical galaxy to us.

utilize a hot plasma to ionize samples efficiently prior to mass spectrometry, which then yields high-precision isotope ratio analysis. These devices have opened new avenues for investigating the composition of rocks and meteorites and thus answer questions about the chronology of Earth and solar system processes. Another modern instrument that permits powerful insights into geochemical and cosmo-chemical processes is the ion microprobe, a mass spectrometer that accepts ions ejected from the surface of a solid sample by the sputtering action of an incident ion beam. The ion microprobe can detect presolar grains in meteorites and deter-mine their isotopic composition, from which details of the synthesis and evolution of the elements in the grain's parent stars can be inferred.

During the last decade, a variety of planets were discovered orbiting around neighboring stars. As more extrasolar planets are discovered and studied, will they help us understand the early history of the solar system? What will they reveal about their own formation? Are there any Earthlike planets around other stars in our neighborhood of the Milky Way? Besides looking for planets, Carnegie scientists are interested in the circumstellar disks that provide the raw materials for planet formation. They are devising sophisticated computer models that will predict the properties of the planets and their atmospheres in the hope of understanding the circumstances under which the extrasolar planetary sys-tems might form.

Earth Science

Earth, our own spaceship, remains for many the most interesting object in the universe. Among the mysteries about Earth is the nature of the materials in the center, the core of the planet. Such questions provided the initial motivation for Carnegie research on materials at high pressure. That goal remains, but the diamond pressure cell can also be used for fundamental studies of various highly compressed molecules and atoms. With increasing pressure, materials undergo a series of transitions in which atoms interact strongly and exhibit unusual properties. For example, elemental sulfur becomes a superconductor near 100 GPa (100 gigapacals equals 1 megabar). What other remarkable changes occur at the pressures that exist in the centers of planets? What useful new materials can be synthesized at pressure? When the sector being built by the Institution at the new Advanced Photon Source at Argonne National Laboratory is completed, the opportunities for discovery will expand because the energies available substantially exceed those produced at older synchrotrons.

Earth is not now and never was static. Seismology provides new insights into Earth's continuing evolution and Carnegie scientists remain dedicated to the development and acquisition of innovative instrumentation. The Institution's 25 portable broadband seismometers are deployed at varying sites around the world and, together with complementary geochemical findings, have revealed that the roots of the most ancient continental crust were stabilized at approximately the same time that the continental crust was put in place.

Can we learn how and even when dynamic processes cause major events in Earth's crust and mantle? Permanent networks of Carnegie strainmeters in

The submersible Alvin, *with a Geophysical Laboratory scientist onboard, explores a hydrothermal vent off the western coast of North America to measure hydrogen levels.*

Residual topography: Seismologists study how forces deep within Earth affect near-surface phenomena.

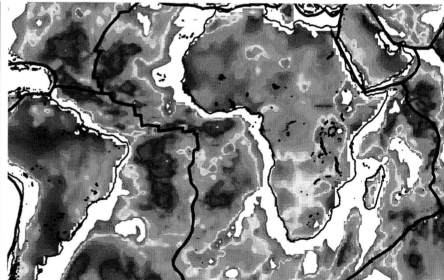

California, Japan, and Iceland provide unique data on crustal deformation and are pointing to ways to predict volcanic eruptions. In February of 2000, for example, colleagues who monitor the strainmeter data at the Icelandic Meteorological Office noted a pattern remarkably similar to one associated with a volcanic eruption almost 10 years earlier and then predicted volcanic activity 20 minutes in advance of the surface eruption. In 1999 two strainmeters of a new design were deployed in deep-sea drill holes about 80 km landward of the Japan Trench. Carnegie strainmeters along with seismometers and global positioning satellite receivers are being installed at a number of locations along the northern San Andreas and Hayward faults in the San Francisco Bay area. This network will serve as a pilot experiment for a more ambitious network that will span much of western North America—the so-called Plate Boundary Observatory.

Water is the single most powerful agent for geologic change on our planet, and it may play or have played an important role on Mars, Europa, and elsewhere. Water reacts chemically with rocks, altering their ratios of oxygen and hydrogen isotopes. Seawater, rainwater, glacier ice, groundwater, and extraterrestrial water each have distinctive isotope ratios. With such data available, questions about where the water originated and even about ancient climate can be answered. Recent age determinations combined with oxygen isotope analyses, for example, demonstrate that a cold climate existed 750 million years ago, coincident with glacial deposits of the "snowball Earth" era. Carnegie scientists have been perfecting *in situ* microanalysis using lasers to analyze oxygen isotopes in silicate and oxide minerals and sulfur isotopes in sulfide minerals to understand the cycling of rocks between Earth's surface and the mantle below.

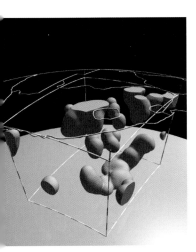

Computer-generated three-dimensional image of seismic velocity anomalies in Earth's mantle beneath southern Africa.

Life Sciences

Fossils were the traditional way to learn about ancient life on Earth. Then, in the 1950s, when fossil amino acids were found in rocks, it became possible to think about chemical fossils. Passage of atoms and molecules through living things leaves traces in the isotopic composition of common elements like oxygen, carbon, nitrogen, and sulfur that yield information about how organisms interacted with their environment. Long after an organism has perished, the abundance of particular isotopes in its residues are discernible. What kinds of samples and from what sites should be examined for biological markers? Besides

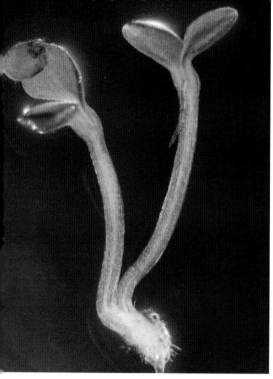
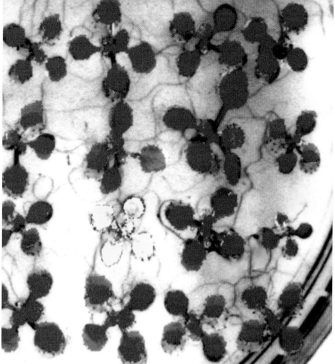
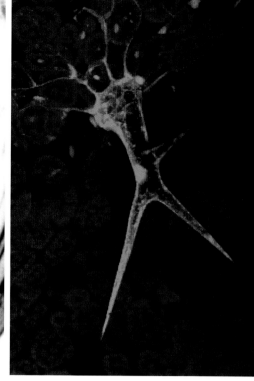

Above: The mustard-like weed Arabidopsis *has a rapid 6-week life cycle that is ideal for studying development, photosynthesis, defenses against disease, and other plant behaviors. In 2000, an international team of scientists, including Carnegie plant biologists, completed the* Arabidopsis *genome—the first for any plant.*

the places already known to be rich in anatomic fossils, new niches now need to be explored. Life on Earth is more robust and pervasive than anyone imagined even a decade ago. There are living things in deep rocks, at hot hydrothermal deep-sea vents, and in the atmosphere, as well as on Earth's surface. The search for markers of early life on Earth has three interrelated goals: to understand how living things contributed to the evolution of the planet, to understand how living things evolved, and to understand how life started. All together, these goals come under the term "geobiology." Some hypothesize that life began in the deep sea in niches like the hot hydrothermal vents. Carnegie scientists are trying to learn how the organic molecules (those that contain carbon) typical of living things might have been formed and stabilized in such places. What role could be played by mineral surfaces in stimulating the synthesis of such molecules?

Geobiology provides clues for where to search for life elsewhere in the solar system or eventually in our galaxy—a newly emerging field called astrobiology. One important clue is the presence of water, which made terrestrial life possible. Thus it makes sense to look for extraterrestrial life where there is or once was water. The search for water on other planets such as Mars and on Europa, a moon of Jupiter, is informed by what is known about the traces that it leaves on Earth and is an integral part of the search for life in those distant places. Designing instruments that can detect other telltale signs of life in such distant places is a challenge. Some day, scientists hope to send a spacecraft carrying such instruments to Mars and even to collect Martian soil samples and return them to Earth.

In one sense, geobiology is about the ecology of the early Earth. Work on the ecology of today's planet is, like geobiology, inherently cross-disciplinary.

Middle: Scientists study Arabidopsis *mutants to understand the molecular workings of the photo-protective mechanism. In this false-color image, a mutant plant (npq1) appears yellow in a field of red, non-mutant plants.*

Right: A trichome (an epidermal hair structure on a plant) appears green when injected with "green fluorescent protein" from a jellyfish. This technique allows scientists to visualize live plant cells and cell components.

Whereas traditional ecology focuses on interactions in small niches, it is now possible to ask questions about the biological and physical mechanisms that control patterns of life and change on a global scale. To mark the beginning of the Institution's second century, Carnegie will establish a new department, to be called Global Ecology, to take advantage of this opportunity. Scientists at the new department will apply satellite observations, computer models, global informatics, and molecular biology to study the large-scale interactions of the biota with the physical properties of Earth.

Ecological studies are informed by knowledge about individual organisms, microbes, plants, and animals. In the last few years, that knowledge has been profoundly deepened by the determination of the sequence of all the DNA—the genomes—of organisms that serve as models for biological research. Included are organisms from bacteria, archaebacteria, and cyanobacteria through humans. An animal or plant is developed as a model for similar organisms because it breeds rapidly and can be experimentally manipulated without a great deal of trouble or expense. Morgan led the way by developing *Drosophila* as a model for animal genetics. Mice serve as a model for mammals, including humans. The short, six-week life cycle and small genome of the tiny *Arabidopsis* plant make it an ideal model for answering questions about the development, photosynthesis, defense against disease, and behavior of green plants. Carnegie scientists played leading roles in the international collaborative efforts that successfully sequenced the genomes of *Drosophila* and *Arabidopsis thaliana*. Carnegie scientists are also active in providing the international scientific community with new technologies and services that make the millions of DNA sequences and banks of mutant organisms accessible and useful.

In the past, the genes required for a particular cellular function were discovered by identifying mutations that interfere with that function. Then the location on chromosomes and the molecular identity of the altered gene must be determined. Today, thanks to the genome data, the approach can be reversed. Genes are identified from their DNA sequence, and the challenge is to determine their function. One option is to alter the gene sequence experimentally so that its function is eliminated or otherwise controllable and to observe the effects; for example, interrupting the gene sequence by insertion of a transposable element, a method that was pioneered by Carnegie biologists almost 20 years ago. Now, thanks to a Carnegie discovery, there is a new way to eliminate the function of particular genes, a technique called RNA interference, or RNAi.

The gene is inactivated by introducing (into certain organisms or cells) a double-stranded RNA complementary to the gene coding sequence. In this way, one or even multiple genes of, for example, flies or worms, can be shut off at precise times during development and the effect on the organism observed.

Carnegie biologists and physical scientists understand that the introduction of new methods and technology is a critical contribution to the advancement of research. Nevertheless, the primary goal of Carnegie biologists is to understand key processes such as development, growth, and response to environmental stimuli. Knowing a genome sequence does not in itself promote such understanding, but because it can help make questions about biological systems more precise, it is substantially useful information for current research.

Carnegie biologists are applying all the modern techniques of genetics, biochemistry, and cell biology to questions about animal and plant development. How are the structures and patterns—the skeletons, nerves, and muscles—of animal bodies, including those of worms, flies, and mice, laid down in very early development? What makes some plant cells become shoots and others roots? Which genes are important for which structures and patterns? Which developmental events are generated within individual cells operating autonomously and which depend on the interactions between cells? What mechanisms ensure that germ cells, those that will give rise to eggs and sperm, are differentiated and sequestered early in animal development? Why, in contrast, can many kinds of plant cells develop into eggs and pollen? How do genes and the proteins they encode build the functional machines that ensure timely and accurate division of cells for growth?

Another set of questions has to do with how organisms respond to external stimuli. How do plants defend themselves from attack by pathogenic viruses, fungi, and bacteria? What mediates the responses of plants, and even animals, to light? Approximately 100 years after Charles and Francis Darwin discovered the phenomenon, Carnegie investigators identified and cloned a gene that enables the model plant *Arabidopsis* to bend toward blue light. They described the chemical mechanism by which the novel protein encoded by this gene responds to the light and initiates the plant's response. Like green plants, cyanobacteria too respond to light in many ways besides photosynthesis. These unicellular organisms contain a protein photoreceptor that is, surprisingly, structurally related to a photoreceptor found in green plants. When this protein is stimulated by light, the cyanobacteria move toward the light source. Mammals too respond to light, as is evidenced by cycles of sleep and wakefulness timed to the rising and

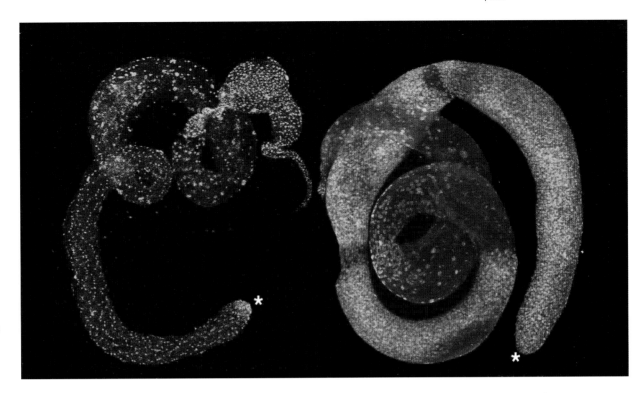

Photomicrograph of two Drosophila *testes, which are used to study some aspects of development. The stars denote the end of the testes.*

setting of the sun. This circadian rhythm reflects an off–on cycle for specific genes in a central control site, the tiny pineal gland.

Although the emphasis of Carnegie science is on advancing knowledge, scientists are also interested in how that knowledge can, in Andrew Carnegie's words, be applied to the improvement of humankind. This goal is evident, for example, in work on the prediction of volcanic eruptions and earthquakes, the investigation of the novel properties of materials at high pressure, the genetic alteration of plants to improve production of food and materials, and the study of early embryonic development.

Scientific discoveries cannot be planned. As in the past, they will arise unexpectedly when individuals can follow the imperatives of personal curiosity. Nevertheless, progress is also made by programs with defined goals. Carnegie scientists will continue to contribute to large-scale collaborative projects like the Messenger mission to Mercury; the new international 2010 Project, whose goal is to determine the function of all *Arabidopsis* genes by 2010; the planning for the next generation of telescopes; and the Plate Boundary Observatory. Still, if the future is to match or surpass the past, the independence and originality of Carnegie scientists must be sustained, remembering always, to paraphrase one of the newest Carnegie staff members, that scientists build instruments and do experiments not because of what we know but because of what we don't know.

The rotunda of the administration building, 1998.

REFERENCES

A decade ago, Michael Aaron Dennis wrote that the Carnegie Institution of Washington was "one of the most important and least studied institutions in the history of American science." (*A Change of State: The Political Cultures of Technical Practice at the MIT Instrumentation Laboratory and the Johns Hopkins University Applied Physics Laboratory, 1930-1945;* Ph.D. dissertation: Johns Hopkins University, 1990, p. 136). Years later, his statement remains largely true so far as comprehensive studies go. There have, however, been many fine publications on particular aspects of the Institution's history. Many of these works were consulted during the writing of this book.

The Institution's *Year Books,* as well as its voluminous archival records, provided content for nearly every chapter of this book. In addition, the National Academy of Sciences' *Biographical Memoirs* series filled in many gaps.

The bibliographic information that follows is intended to guide the interested reader to sources that were examined during the preparation of this book.

NETTIE STEVENS
Cross, Patricia C. and John P. Steward. 1993. "Nettie Maria Stevens: Turn-of-the-century Stanford alumna paved path for women in biology," *Sandstone & Tile,* 17 (1): 3-12.

ANDREW CARNEGIE
Carnegie, A. 1889. "Wealth," *North American Review,* 148 (June, 1889): 653-664; and 149 (December, 1889): 682-698. Later editions were published by various companies in book form under the title, *The Gospel of Wealth.*

Carnegie, A. 1920. *Autobiography of Andrew Carnegie.* Boston, [no publisher listed], 1920.

Gregorian, V. 2000. *Report of the President.* New York: Carnegie Corporation of New York. The actual quote reads: "When it comes to reinventors and reshapers of American society, Andrew Carnegie personifies the difference an individual with an altruistic vision can make."

Hendrick, B. 1932. *The Life of Andrew Carnegie,* 2 vols. Garden City, New York: Doubleday, Doran.

Lester, R. M. 1941. *Forty Years of Carnegie Giving: A Summary of the Benefactions of Andrew Carnegie and of the Work of the Philanthropic Trusts.* New York: Charles Scribner's Sons.

Maurer, H. 1948. "Andrew Carnegie," *Fortune,* 37: 116-148.

Schneiderman, H. A. 1985. "What biotechnology has in store for us," *Carnegie Evening, 1985.*

Wall, J. F. 1970. *Andrew Carnegie.* New York: Oxford University Press.

CHARTING THE COURSE
Works describing the founding of the Carnegie Institution as well as works covering the broader picture of American philanthropy were consulted, including:

Hanson, E. 2000. *Achievements: The Rockefeller University: A Century of Science for the Benefit of Humankind, 1901-2001.* New York: Rockefeller University Press.

Kohler, R. E. 1991. *Partners in Science: Foundations and Natural Scientists, 1900-1945.* Chicago: University of Chicago Press.

Lagemann, E. 1983. *Private Power for the Public Good: A History of the Carnegie Foundation for the Advancement of Teaching.* Middletown, Conn.: Wesleyan University Press.

Madsen, D. 1969. "Daniel Coit Gilman at the Carnegie Institution of Washington," *History of Education Quarterly,* v. 9 (2): 154-186.

Miller, H. S. 1970. *Dollars for Research: Science and Its Patrons in Nineteenth-Century America.* Seattle: University of Washington Press.

Reingold, N. 1979. "National science policy in a private foundation: The Carnegie Institution of Washington," in A. Oleson and J. Voss, eds., *The Organization of Knowledge in Modern America, 1860-1920.* Baltimore: Johns Hopkins University Press.

Wooster, M. M. 2001. "A philanthropist at work: The letters of Andrew Carnegie," *Philanthropy,* 15 (3): 24-28.

Yochelson, E. L. 1994. "Andrew Carnegie and Charles Doolittle Walcott: The origin and early years of the Carnegie Institution of Washington," in G. A. Good, ed., *The Earth, the Heavens, and the Carnegie Institution of Washington.* Washington, D.C.: American Geophysical Union.

Yochelson, E. L. 1998. *Charles Doolittle Walcott, Paleontologist.* Kent, Ohio: Kent State University Press.

THE DEPARTMENTS
Boss, B. 1968. *History of the Dudley Observatory, 1852-1956.* Albany, N. Y.: Dudley Observatory.

Good, G. A., ed. 1994. *The Earth, the Heavens and the Carnegie Institution of Washington.* Washington, D.C.: American Geophysical Union.

Haskins, C. P., ed. 1967. *The Search for Understanding: Selected Writings of Scientists of the Carnegie Institution, Published on the Sixty-Fifth Anniversary of the Institution's Founding.* Washington, D.C.: Carnegie Institution of Washington.

Hickman, C. M. 2001. "Building for science: Carnegie Institution of Washington's Geophysical Laboratory," *Washington History,* 13:1 (spring/summer), 32-51.

Kevles, D. J. 1985. *In the Name of Eugenics: Genetics and the Uses of Human Heredity.* New York: Alfred A. Knopf.

Kevles, D. J. 1997. *The Physicists: The History of a Scientific Community in Modern America.* Cambridge, Mass.: Harvard University Press.

Lankford, J. 1987. "Charting the Southern Sky," *Sky & Telescope.* September: 243-246.

Servos, J. W. 1983. "To explore the borderland: The foundation of the Geophysical Laboratory of the Carnegie Institution of Washington," *Historical Studies in the Physical Sciences,* 14: 147-185.

Yochelson, E. L. and H. S. Yoder, Jr. 1994. "Founding the Geophysical Laboratory, 1901-1905: A scientific bonanza from perception and persistence," *Geological Society of America Bulletin,* 106: 338-350.

Yoder, Hatten S., Jr. 1994. "Development and promotion of the initial scientific program for the Geophysical Laboratory," in Good, *The Earth, the Heavens and the Carnegie Institution of Washington,* 22-28.

THE WAR YEARS
This chapter benefited from Louis Brown's manuscript history of the Department of Terrestrial Magnetism, which will be published in 2002. The following works were also helpful:

Bush, V. 1949. *Modern Arms and Free Men: A Discussion of the Role of Science in Preserving Democracy.* New York: Simon and Schuster.

Bush, V. 1970. *Pieces of the Action.* New York: William Morrow and Company.

Bush, V. 1990 (originally published in 1945). *Science—The Endless Frontier: A Report to the President on a Program.* Washington, D.C.: National Science Foundation.

Doel, R. E. 1996. *Solar System Astronomy in America: Communities, Patronage, and Interdisciplinary Research, 1920-1960.* Cambridge: Cambridge University Press.

Zachary, G. P. 1997. *Endless Frontier: Vannevar Bush, Engineer of the American Century.* New York: The Free Press.

VISION AND REVISION
This chapter benefited from conversations with Robert E. Kohler, Louis Brown, Hatten S. Yoder, Jr., and Philip Abelson. See also:

De Solla Price, D. J. 1986. *Little Science, Big Science and Beyond.* New York: Columbia University Press.

Dyson, F. 1992. *From Eros to Gaia.* New York: Pantheon.

Galison, P. and B. Hevly, eds. 1992. *Big Science: The Growth of Large-Scale Research.* Stanford, Calif.: Stanford University Press.

Haskins, C. P. 1994. *This Our Golden Age: Selected Annual Essays of Caryl P. Haskins, President, Carnegie Institution of Washington, 1956-1971;* ed. by J. D. Ebert. Washington, D.C.: Carnegie Institution of Washington.

Tuve, M. A. 1959. "Is science too big for the scientist?" *Saturday Review,* June 6, 1959: 48-52.

BUILDING TELESCOPES
Adams, W. S. 1947. "Early days at Mount Wilson," *Publications of the Astronomical Society of the Pacific,* 59 (October): 213-231; (December): 285-304.

Osterbrock, D. E. 1993. *Pauper and Prince: Ritchey, Hale & Big American Telescopes.* Tucson: University of Arizona Press.

Wright, H. 1966. *Explorer of the Universe: A Biography of George Ellery Hale.* New York: E. P. Dutton & Co.

Wright, H., J. N. Warnow, and C. Weiner, eds. 1972. *The Legacy of George Ellery Hale: Evolution of Astronomy and Scientific Institutions, in Pictures and Documents.* Cambridge: MIT Press. See especially "Hale and the role of central scientific institutions in the United States," pp. 273-283, by D. J. Kevles.

THE FABRIC OF THE UNIVERSE

Baade, W. 1963. *Evolution of Stars and Galaxies.* Ed. by C. Payne-Gaposchkin. Cambridge: Harvard University Press.

Christianson, G. E. 1995. *Edwin Hubble: Mariner of the Nebulae.* New York: Farrar, Straus and Giroux.

Hubble, E. 1929. "Exploration of space," *Harper's Magazine,* 158 (May): 732-738.

Hubble, E. 1936. *The Realm of the Nebulae.* New Haven: Yale University Press.

Osterbrock, D., R. Brashear, and J. Gwinn. 1990. "Young Edwin Hubble," *Mercury,* January/February: 3-14.

DARK MATTER

This chapter is based on an interview with Vera Rubin. Portions of the text benefited from R. M. Hazen and M. Singer. 1997. *Why Aren't Black Holes Black?* New York: Anchor Books/ Doubleday. Other useful references were:

Bartusiak, M. 1990. "The woman who spins the stars," *Discover* (October): 88-94.

Berendzen, R. et al. 1976. *Man Discovers Galaxies.* New York: Science History Publications.

Rubin, V. 1997. *Bright Galaxies, Dark Matters.* Woodbury, N.Y.: American Institute of Physics.

Rubin, V. C., W. K. Ford, Jr., and N. Thonnard. 1978. "Extended rotation curves of high-luminosity spiral galaxies. IV. Systematic dynamical properties, Sa→Sc," *The Astrophysical Journal,* 225 (November 1): L107-L112.

TERRESTRIAL MAGNETISM

This chapter relied heavily on conversations with Shaun Hardy and on Louis Brown's manuscript history of the Department of Terrestrial Magnetism, which will be published in 2002 Also useful were:

Cooper, D. 1998. "Surveying the Pacific: The voyages of the *Galilee,* 1905-1908," *Mains'l Haul* (winter): 36-47.

Good, G. A. ed. 1994. *The Earth, the Heavens and the Carnegie Institution of Washington.* Washington, D.C.: American Geophysical Union.

Multhauf, R. P. and G. Good. 1987. *A Brief History of Geomagnetism and A Catalog of the Collections of the National Museum of American History.* Washington, D.C.: Smithsonian Institution Press.

SEISMOLOGY

Geschwind, C.-H. 2001. *California Earthquakes: Science, Risk and the Politics of Hazard Mitigation.* Baltimore: Johns Hopkins University Press.

Goodstein, J. R. 1984. "Waves in the Earth: Seismology comes to Southern California," *Historical Studies in the Physical Sciences,* 14: 201-230.

Goodstein, J. R. 1991. *Millikan's School: A History of the California Institute of Technology.* New York: W. W. Norton & Co.

MEGABAR

This chapter drew heavily on R. M. Hazen, 1993. *The New Alchemists: Breaking Through the Barriers of High Pressure.* New York: Times Books. Additional information on the subject of high-pressure research is found in R. M. Hazen. 1999. *The Diamond Makers.* Cambridge: Cambridge University Press.

HEREDITY

This chapter benefited from conversations with Joseph Gall, Patricia Gossel, and Maxine Singer. The following works were also consulted:

Fedoroff, N. and D. Botstein, eds. 1992. *The Dynamic Genome: Barbara McClintock's Ideas in the Century of Genetics.* Cold Spring Harbor, N.Y.: Cold Spring Harbor Laboratory Press.

Keller, E. F. 1983. *A Feeling for the Organism: The Life and Work of Barbara McClintock.* San Francisco: W. H. Freeman.

Kohler, R. E. 1994. *Lords of the Fly:* Drosophila *Genetics and the Experimental Life.* Chicago: University of Chicago Press.

Micklos, D. 1988. *The First Hundred Years: A History of Man and Science at Cold Spring Harbor.* Cold Spring Harbor, N.Y.: Cold Spring Harbor Laboratory Press.

Stahl, F. W. 2000. *We Can Sleep Later: Alfred D. Hershey and the Origins of Molecular Biology.* Cold Spring Harbor, N. Y.: Cold Spring Harbor Laboratory Press.

Witkowski, J. 2000. *Illuminating Life: Selected Papers from Cold Spring Harbor, 1903-1969.* Cold Spring Harbor, N.Y.: Cold Spring Harbor Laboratory Press.

EMBRYOLOGY

Brown, D. D. 1987. "The Department of Embryology of the Carnegie Institution of Washington," *BioEssays* 6: 92-96.

Brown, D. D. 1994. "Some genes were isolated and their structure studied before the recombinant DNA era," *BioEssays* 16: 139-143.

O'Rahilly, R. 1973. *Developmental Stages in Human Embryos; Including a Survey of the Carnegie Collection,* CIW Publication no. 631. Washington, D.C.: Carnegie Institution of Washington.

O'Rahilly, R. 1988. "One hundred years of human embryology," *Issues and Reviews in Terratology,* 4: 81-128.

Ramsey, E. M. 1975. *The Placenta of Laboratory Animals and Man.* New York: Holt, Rinehart, and Winston.

Sabin, F. R. 1934. *Franklin Paine Mall: The Story of a Mind.* Baltimore: The Johns Hopkins University Press.

PHOTOSYNTHESIS

This chapter relied heavily on Patricia Craig's manuscript history of the Carnegie Institution's Department of Plant Biology, which will be published in 2002. Also useful were the following:

Bowers, J. E. 1990. "A debt to the future: Scientific achievements of the Desert Laboratory, Tumamoc Hill, Tucson, Arizona." *Desert Plants,* 10, 1: 9-12, 35-47.

Bowers, R. 1992. *Mr. Carnegie's Plant Biologists: The Ancestry of Carnegie Institution's DPB.* Washington, D.C.: Carnegie Institution of Washington.

Spoehr, H. A. 1926. *Photosynthesis.* New York: Chemical Catalog Company.

ECOLOGY

This chapter was adapted from Patricia Craig's forthcoming book on the history of the Carnegie Institution's Department of Plant Biology. Also useful were:

Bowers, J. E. 1986. "A career of her own: Edith Shreve at the Desert Laboratory," *Desert Plants,* 8: 23-29.

Bowers, J. E. 1988. *A Sense of Place: The Life and Work of Forrest Shreve.* Tucson: University of Arizona Press.

Bowers, R. 1992. *Mr. Carnegie's Plant Biologists: The Ancestry of Carnegie Institution's DPB.* Washington, D.C.: Carnegie Institution of Washington.

Britton, N. C. and J. N. Rose. 1920. *The Cactaceae: Descriptions and Illustrations of Plants of the Cactus Family,* 4 vols. Washington, D.C.: Carnegie Institution of Washington.

Clements, E. 1960. *Adventures in Ecology: Half a Million Miles.* New York: Pageant Press.

Hornaday, W. T. 1914. *Camp-Fires on Desert and Lava.* New York: 1914.

McGinnies, W. G. 1981. *Discovering the Desert: Legacy of the Carnegie Desert Botanical Laboratory.* Tucson: University of Arizona Press.

ARCHEOLOGY

Bishop, R. L. and F. W. Lange, eds. 1991. *The Ceramic Legacy of Anna O. Shepard.* Niwot: University Press of Colorado.

Brunhouse, R. L. 1971. *Sylvanus G. Morley and the World of the Ancient Mayas.* Norman: University of Oklahoma Press.

Deuel, L. 1967. *Conquistadors Without Swords: Archaeologists in the Americas; An Account with Original Narratives.* New York: St. Martin's Press.

Givens, D. R. 1992. *Alfred Vincent Kidder and the Development of Americanist Archaeology.* Albuquerque: University of New Mexico Press.

Sabloff, J. A. 1990. *The New Archaeology and the Ancient Maya.* New York: Scientific American Library.

Woodbury, R. B. 1973. *Alfred V. Kidder.* New York: Columbia University Press.

BIOPHYSICS

This chapter relied heavily on Louis Brown's history of the Department of Terrestrial Magnetism, to be published in 2002. Carnegie publications, including the *Year Books,* added to the content.

INDEX